THE GOOD HOPE CANNERY

THE GOOD HOPE CANNERY

Life and Death at a Salmon Cannery

W.B. MacDonald

The Good Hope Cannery in 2009.

Page 2: Good Hope Cannery, c. 1900. Photo courtesy of Ian Bell-Irving.

IN MEMORY OF GUIDE AARON BEECHING

"Tip up! Tip up! Reel! Reel! Reel!"

Women hand packing the salmon into tins at Wadham's cannery, Rivers Inlet, c. 1900. Photo courtesy of Ian Bell-Irving.

Copyright © 2011 Bruce MacDonald
01 02 03 04 05 06 16 15 14 13 12 11

All rights reserved. No part of this publication may be reproduced, stored in a retrieval system or transmitted, in any form or by any means, without prior permission of the publisher or, in the case of photocopying or other reprographic copying, a licence from Access Copyright, the Canadian Copyright Licensing Agency, www.accesscopyright.ca, 1-800-893-5777, info@accesscopyright.ca.

Edited by Patricia Wolfe.
Text design by Vici Johnstone.
Cover design by Pamela Cambiazo.
Printed in Canada.

Caitlin Press Inc.
8100 Alderwood Road,
Halfmoon Bay, BC V0N 1Y1
www.caitlin-press.com

Caitlin Press Inc. acknowledges financial support from the Government of Canada through the Canada Book Fund and the Canada Council for the Arts, and from the Province of British Columbia through the British Columbia Arts Council and the Book Publisher's Tax Credit.

 Canada Council for the Arts / Conseil des Arts du Canada

 BRITISH COLUMBIA ARTS COUNCIL

Library and Archives Canada Cataloguing in Publication

MacDonald, Bruce, 1956-
    The Good Hope Cannery : life and death at a salmon cannery / W.B. MacDonald.

Previously published under title: Good Hope.
Includes bibliographical references.
ISBN 978-1-894759-64-9

    1. Good Hope Cannery—History. 2. Salmon canneries—British Columbia—Rivers Inlet—History. 3. Fishing lodges—British Columbia—Rivers Inlet—History. 4. Rivers Inlet (B.C.)—History. 5. Rivers Inlet (B.C.)—Biography. I. Title.

HD9330.S33G66 2011   338.3'72756097111   C2011-904939-2

Contents

Prologue	11
Keeping the Books	15
Men of Good Hope	23
The Letter Book	43
Catch What You Can, Can What You Catch	55
Marea's Story	93
Square Hooks	109
The Storekeepers	133
The Anderson Years	145
Hopeful	161
Self-propelled Canoes	187
Epilogue	210
Acknowledgements	211
Bibliography	212
Sources	214
Personal Interviews	215

The village of Oowekeeno, Rivers Inlet, British Columbia, early 1900s. Photo courtesy of the United Church of Canada.

Prologue

In February 2008, Tony Allard, owner of the Good Hope Cannery, hired me to write its history. At our initial meeting Allard told me that not many historical artifacts seemed to have survived, although there were some bookkeeping records, somewhere. I asked if I could I see them, and he said, of course, he'd have someone find them and ship them down. A number of weeks passed, during which time a search was conducted. Finally an email went out to a former employee of the cannery who had been in charge of storing the records. Her reply provided me with the perfect opening to the story of Good Hope.

Good Hope Cannery was founded in 1895 by Scottish entrepreneur, engineer, and outdoor adventurer Henry Ogle Bell-Irving. Good Hope canned salmon continuously until 1940 and thereafter served company fishermen as a place where they could refuel, eat, buy supplies and have their boats and nets repaired. By the late 1960s depleted fish stocks and technological advances rendered Good Hope obsolete as a camp. But a Henry Bell-Irving descendant, grandson Ian Bell-Irving, envisioned Good Hope as a sport fishing resort catering to affluent North Americans, and so Good Hope entered the third phase of its life, a life that continues to this day. The Good Hope Cannery and the Goose Bay Cannery is all that are left of an important era in BC's history—all the other canneries in Rivers Inlet have vanished

The Good Hope Cannery is not hard to find, provided you can find the inlet named by Captain George Vancouver in honour of George Pitt, First Baron Rivers. Rivers Inlet is

THE GOOD HOPE CANNERY

For Britons, if the meat in the can wasn't red, it wasn't salmon. An Anglo-British Columbia Packing Company Red King poster, c. 1915. Image courtesy of the City of Vancouver Archives.

372 miles (595 kilometres) by air from Vancouver, British Columbia. A fjord in the Central Coast region of the province, the inlet is accessible only by boat or seaplane, its entrance from Dean Channel about 78 miles (125 kilometres) southwest of Bella Coola and about 40 miles (64 kilometres) north of the northern tip of Vancouver Island. The inlet is 28 miles (45 kilometres) long from its head at the community of Rivers Inlet, home to the Wuikinuxv people, also known as the Oowekeeno or Rivers Inlet people. The Oowekeeno have lived for millennia along the Wannock River, the inlet's primary waterway, feeding in from Owikeno Lake, a fresh-water body about 31 miles (50 kilometres) long. It is the spawning ground tributaries of Owikeno Lake that returning salmon seek, and there hangs the history of the Good Hope Cannery.

The cannery is on the west side of the inlet, more or less equidistant from the inlet's mouth and head. It is sheltered within a hook-shaped bay whose mouth opens to the south. Three "sister" islands—Ida, Ethel and Florence—extend south from the tip of the "hook." Ida, the smallest, is closest to the tip followed by slender Ethel and curvaceous Florence. Sandell Lake, named after Good Hope watchman Olof Sandell, is in the mountains high above and 2.5 miles (4 kilometres) west of Good Hope. As the Sandell River, it spills forcefully—and icily—into the inlet, just a few hundred metres south of the cannery. A sheltered cove and a steady source of fresh water (said to be the best in the inlet) account for Good Hope's location, but the comforting providence of the three lovely "sisters" doesn't hurt either.

Seventy-five years passed from the first salmon canned in Rivers Inlet, at Rivers Inlet Cannery (RIC) in 1882, to the last, in 1957 at Goose Bay. RIC was at the head of the inlet, Goose Bay at the mouth, creating a nice symmetry. Good Hope canned its last salmon even earlier,

12

PROLOGUE

in 1940. Advances in transportation technology, enabling fish processing to be done near urban centres, rendered the canneries obsolete. A few, such as Wadham's and Good Hope, survived their obsolescence by servicing fishermen and their boats. But this phase was much shorter lived. In 1970 Good Hope, hoping to live up to its name, was reinvented by Ian Bell-Irving, grandson of Good Hope's founder Henry Ogle Bell-Irving, as a sport fishing lodge, the inlet's first. The strategy worked. And because it did, Good Hope escaped the fate of all the other Rivers Inlet canneries save Goose Bay.

Today next to nothing remains of the other canneries—Beaver, Green's, Victoria and the rest. It's hard to imagine

In the days before gas-engined fishing boats, fishermen used skiffs equipped with oars and sails. Good Hope Cannery is visible in the middle distance in this photo taken in 1895 by H.O. Bell-Irving. Photo courtesy of Patricia Wilson.

they ever existed. To the keen eye a few pilings visible at low tide give a hint that something was once there. Closer examination of a beach might reveal the odd red brick or two. Venturing a few steps into the forest you might stumble over a cast iron grate, a rusty bucket minus its bottom, or a length of broken water pipe. Going even farther, usually up a steep embankment, you might come upon a cannery garbage dump, its "treasure" of broken bottles, cracked plates and corroded tin scraps nestled among conifer needles, salal and moss. But speeding up the inlet on the way to catch a "big spring," your eyes fixed on the distance, there's no way of telling that more than a dozen canneries with all their ancillary buildings, wharves and floats, countless people, and, at the peak, a thousand gillnet boats a season, once occupied Rivers Inlet.

In 1980 disgruntled employees took out their frustration on an already worn-down Good Hope. The office is on the left and the cannery is on the right. Photo courtesy of Darion Jones.

Chapter 1

Keeping the Books

The situation of the Good Hope cannery is poor, but the cannery itself is well appointed.
—Henry Doyle, 1902
First manager of the British Columbia Packers' Association. (1874–1961)

In October 2006 a crew was demolishing an old, one-storey building at the Good Hope Cannery in Rivers Inlet, British Columbia. The building was known to Good Hope staff as "the Hilton" because for many years it provided accommodation to guides and male lodge workers. But, built in 1895, it was certainly no Hilton. Decades of exposure to pounding Central Coast rainstorms, heavy snows, driving winds and salt air had taken its toll. Its foundations—pilings—were rotten, the roof leaked, the floor and walls tilted at weird angles, and it was full of mould.

Tearing down an inside wall, the bemused crew discovered stacked between the studs seven cardboard cases 12 inches long by 10 inches high and 4 inches wide (30 x 25 x 10 centimetres). Given the draftiness of these old cannery buildings and the stormy coastal weather, someone had probably stashed these cases to provide a small measure of insulation. But if that were true, why weren't there lots of other cases stacked between the studs?

One of the boxes contained bookkeeping records—tally sheets and receipts—for the 1943 fishing season. The papers were in mint condition, stacked neatly in chronological order.

They were held in place by two metal rings, now corroded but still serving their purpose. The one-page tally sheets recorded each fisherman's credits and debits down to the penny. That year sockeye was worth 14 cents a pound while red spring fetched 7 cents. The receipts were for cheques made out to fishermen by the Anglo-British Columbia Packing Company. The name meant nothing to the demolition crew, and why should it? The company—ABC as it was known—hadn't existed for well over thirty years. As they flipped through the pages the fishermen's names didn't register with any of them. Long dead, they thought. One guy had a good season, over a thousand sockeye totalling 6,049 pounds (2,750 kilograms), equalling $846.86—not bad at all. The crew would have spent more time poring over the records but they had a job to do in 2006, and 1943 was over and done with. They slid the paperwork back into its jacket and took the seven cases to the cannery office for safekeeping.

Eventually the cases, along with other cannery artifacts, were shipped by air to Good Hope's office in Richmond where Morag Wehrle, an archivist under contract with Good Hope, examined and catalogued the material. It then travelled back to Good Hope. Many of the artifacts went on display in the cannery, but what to do with the bookkeeping records? In the meantime a new building had taken the place of the Hilton; the finishing touches were being put on its interior. Candace Meagher, a Simon Fraser University history student on Good Hope's staff, had the perfect spot in mind for them—in the wall of the new Hilton. The fishermen of the 1940s went back into hiding, as a measure of safekeeping, there being no other place to put them.

In spring 2008 Mark Dowd, the cannery's general manager, handed me a box containing the seven cases and other material. They had travelled nine hundred miles (1448 kilometres)

Rivers Inlet canneries issued coupons or "scrip" in lieu of cash wages that were redeemable only at the issuing company's store. There was almost nowhere else in the inlet to spend money anyway. Courtesy of the Good Hope Cannery.

16

Rivers Inlet Canneries

Blacksmith Robert Draney and Victoria merchant Thomas Shotbolt built the first of Rivers Inlet's fourteen canneries in 1882 at the head of the inlet, at the mouth of the Wannock River. Rivers Inlet Cannery (RIC) was followed one year later by the Victoria Cannery, also at the mouth of the Wannock, opposite RIC. They were joined in 1884 by the Wannock Cannery at Wannock Cove, on the north shore of the inlet, west of Moses Inlet. Good Hope was built next in 1895. Two years later saw the establishment of three more canneries: Brunswick stood at Canoe Pass, near the Wannock Cannery; Green's (also called Vancouver) was about a quarter mile from Brunswick, across the bay from Wannock; and Wadham's was 2 miles (3.2 kilometres) south of Good Hope. In 1906 three more canneries sprang up: Beaver, in Schooner Passage; Kildala in Kildala Bay; and Strathcona between Wadham's and Good Hope. In 1917 Provincial Cannery joined Beaver in Schooner Passage and then, in 1918, the McTavish Cannery was erected on the south side of the inlet, across from Kildala. Goose Bay Cannery was built near the mouth of the inlet in 1926. The last cannery, Moses Inlet Cannery, went up in 1932, only to close in 1935, a casualty of the Great Depression.

and over sixty years. In addition to the seven cases, the box contained a Good Hope ledger book for 1937–43 and a number of loose papers, including the 1939 business records for the McTavish Cannery. How had records for McTavish, about 10 miles (16 kilometres) farther up Rivers Inlet, gotten mixed in with Good Hope? The Good Hope records covered the 1943, 1944 and 1945 seasons, April through August for the most part. There were receipts, memos, invoices, fishermen's accounts, bills of sale, letters and an envelope containing 1943 Good Hope coupons worth from five cents to a dollar. The fishermen had come from all over British Columbia: New Westminster, Langley, Aldergrove, Vancouver, Sointula, Nanaimo, Kuper Island, Quathiaski Cove, Simoon Sound and Alert Bay. The name that stood out—you couldn't

THE GOOD HOPE CANNERY

it because his signature was everywhere—was that of Good Hope manager Levi Lauritsen. He had been a busy man, arranging boat repairs, advancing or denying money to fishermen, selling boats on consignment and attending to a hundred other details.

Among Lauritsen's correspondents was William Williams, who wrote, "My sons Alfred and Albert expect to get 30 days leave from the Air Force during the summer, and if they get their leaves during July they plan to fish for you in Rivers Inlet." Peder Skarsberg of Sexsmith, Alberta, was also looking for employment: "I got a letter from L. Davidsen who has been working for you before. He got my brother a job with you and I would like to know if you could give me a job too. I have good experience with fishing as I have been fishing in Norway and on the Pacific coast. Please tell me when you start and where I could meet you." Lauritsen wrote him back, "We will reserve fishing gear for you for the coming season."

Other jobseekers were not so lucky. James H. Shaughnessy of Alert Bay wrote to Good Hope's manager, "I'm coming [to] try my luck again this year, and I wonder if you can get the net boss to get me two years old net… and if you can send me $25 advance…" Lauritsen responded, "We are sorry but we can not do anything for you this year."

Then there was L. McNutt, who wondered if it would be all right to have his two large dogs with him while he fished. "I think it would be better for you to stay with the job you have," replied Lauritsen, "rather than come up here and take a chance on what you will earn. Furthermore two large dogs would be a nuisance to yourself and others especially on a small skiff."

Some of the letters were profoundly sad. "I regret to inform you that we have lost our beloved father in a car accident," wrote Peter Good. "He was struck while crossing a street in Seattle. The man who run over him was drunk… I would like to take over where my father left off if it is possible will you write and let me know?" Good wrote again to Lauritsen in March, "If it is possible Mr. Lauritsen I would like to get help to repair and paint the boat. I will have to buy my own tools, & steam box equipment, and also cotton, nails, and so on. I'd say $25 would help to cover half of the cost." Good, on behalf of his friend Moses Wilson, asked Lauritsen, "Is there absolutely no advances this year?" He noted that Moses "is not in debt to the company."

Frank Dawson of Simoon Sound wrote Lauritsen, "Sorry to let you know that my Brother Harry Dawson is died at the Alert Bay Hospital Oct. 4–44. About his gillnet boat…" In April he wrote again, obviously employed. "Will you talk to Ludvig the net boss to look after the girls at their work so they can work same amount of net. Mrs. L. Peters wasn't good for my girls last year."

Lauritsen's character comes through in everything he wrote. "According to the debit note received the *Golden Arrow* must have been a liability to you," he informed Knight Inlet Cannery manager W.J. Matthews. "We delivered her to you last fall without cost and now we are charged with return trip, even with the telegram sent by skipper at your own request. I cannot see that this is playing fair, can you?" He wrote to Eli Sampson, Alex James and Sam Wilson to either deliver fish as agreed or their boat would be seized. In a letter to Burns & Co. in Calgary Lauritsen complained about the poor quality of their beef. "The problem," the company replied, "is in procuring beef of any kind rather than selection." To help Dr. Darby with the running expenses of the Rivers Inlet hospital, he collected two dollars from 143 Good Hope fishermen and matched their donations equally with cannery money. It fell to Lauritsen to inform the same doctor that Werner Smedman "was found dead on his boat this morning."

But of all the documents, the one that most intrigued me was a voters list.

The 1945 document listed twenty voters at Good Hope, most of them fishermen, most from the Lower Mainland, but to five of them Good Hope Cannery was home. Sarah Irene

One of hundreds of pieces of correspondence discovered in bookkeeping binders hidden behind a wall in Good Hope's office. Letter courtesy of Good Hope Cannery.

THE GOOD HOPE CANNERY

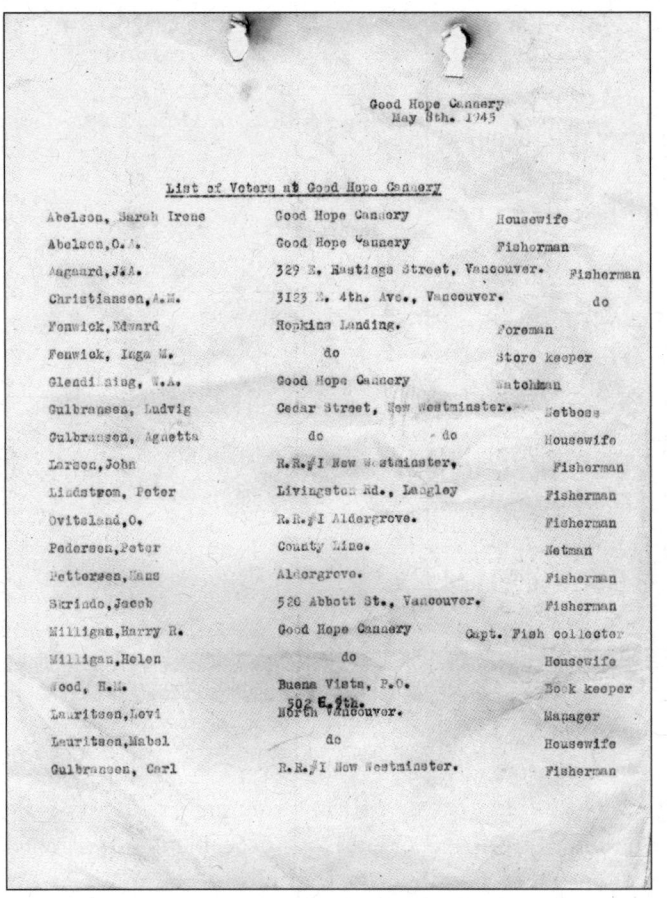

On June 11, 1945, Canadian voters re-elected a Liberal government under Prime Minister William Lyon Mackenzie King for the third consecutive time. Twenty voters in that election were residents of the Good Hope Cannery. Document courtesy of Good Hope Cannery.

Abelson was married to O.A. Abelson, a fisherman. Harry R. Milligan was the captain of one of the cannery's fish collector boats; Helen was his wife. W.A. Glendinning was the cannery's watchman. It was surprising that these people lived year-round at Good Hope. I had assumed that at the end of the fishing season everyone went home and the cannery shut down until the following season. But not so: some hardy souls stayed on through the fall, winter and spring. The Abelsons and the Milligans were married. Glendinning, it seemed a safe bet, was either a bachelor or a widower. Levi Lauritsen of North Vancouver was the cannery's manager; Mabel was his wife. Harold Wood was the bookkeeper. Inga Fenwick minded the store. She was somehow related to foreman Edward Fenwick, possibly his wife. Ludvig Gulbransen of New Westminster was the net boss and Agnetta was his wife. The others were all fishermen: J.A. Aagaard, A.M. Christiansen, Carl Gulbransen, John Larson, Peter Lindstrom, O. Ovitslund, Hans Pedersen and Jacob Skrindo.

As I examined the list I took in all the minor typing mistakes that people were prone to making in the days of typewriters. Certain letters were faint from having been struck with too little pressure, other letters were superimposed forcefully on incorrect strikes. In "Aagaard, J.A." there was a faint shadow of a question mark after the "J" that had been corrected with a heavily hammered period. This document bore many telltale signs that it had been created by human hands. The printed product of a computer bears none of these. This voters list, like the other documents in the seven bookkeeping cases, was obviously, unmistakably once handled by human hands. Who were these people? And what were their lives all about at Good Hope Cannery during the war years? I didn't know and, realistically, I figured I'd never know. The list was over sixty years old and most, if not all of them, had to be long gone. But, as it turned out, they were not all long gone.

Henry Ogle Bell-Irving, engineer, entrepreneur, big game hunter and master of all details, c. 1886. Photo courtesy of Patricia Wilson.

Chapter 2

Men of Good Hope

A dangerous lunatic was brought down from Rivers Inlet, to day, by Constable Loe. His name is McDaniels, and he has followed the calling of a fisherman for some years.
—*Victoria Daily Colonist*, September 11, 1892

Born into a wealthy family in Lockerbie, Scotland, in 1856, Henry Ogle Bell-Irving had been educated in Edinburgh before studying engineering at Germany's prestigious Karlsruhe University. Henry's father had suffered serious business losses and, following his untimely death, the family found itself in straitened financial circumstances. After qualifying as a civil engineer and practising in Europe for four years, Henry arrived in Canada in 1882 determined to rebuild the family fortune. He went to work as a survey engineer for the Canadian Pacific Railway and then, in 1885, entered the real estate business in Vancouver. Shortly after, he established an importing business with R.P. Paterson. The firm of Bell-Irving and Paterson was the first to have general cargo shipped directly from London to Vancouver via Cape Horn. In 1890, Bell-Irving secured options on nine British Columbia fish canneries—seven on the Fraser River and two on the Skeena River—and established the Anglo-British Columbia Packing Company (ABC).

The partners' venture took advantage of legislation enacted in the early 1890s creating the legal format of the limited company. Prior to this legislation there were no laws for limited liability or even for registered partnerships. Businesses, including canneries, were sole proprietorships, associations, partnerships or companies, often started by men who had participated in the Klondike gold rush and were looking for a better bet. These men arranged with a fiscal agent or commission house, usually in Victoria or San Francisco, to import canning machinery and supplies from Britain, and to ship the product to market, again usually Britain. Because it could take up to eighteen months to fill an order, with no guarantee that a shipment would not be lost at sea, cannery men operated on loans advanced by commission houses. It was a risky business for the canner, though it could be a very profitable one for the agent.

With the arrival of a new breed of businessmen the number of canneries and the output of canned salmon jumped to new heights. British Columbia Canning Company was the first to form in 1889. Two years later limited companies such as ABC and the Victoria Canning Company controlled over 60 percent of the Fraser River's pack. These upstarts greatly decreased the importance of local capital, especially that of commission agents. BC Packers was backed by a consortium of eastern Canadian financial interests, and ABC was backed in the United Kingdom. On December 22, 1890, the Anglo-British Columbia Packing Company was incorporated in Middlesex, England. Bell-Irving and Paterson were appointed managing and selling agents. As such, they controlled the insurance, received a 2.5 percent commission on purchases, and got 5 percent for selling the pack. They maintained a head office for the company in Vancouver, and titular head office in London, England.

In July 1891 Bell-Irving was in Rivers Inlet scouting a location for an ABC cannery. Early in 1892 he made a decision not to build, at least, not yet. In December 1893, as he sailed for London and the annual meeting of the ABC company (he would not become a director of the company until the 1920s), he read in a book a philosophy of life that matched his own:

1. Have a definite aim.
2. Go straight for it.
3. Master all details.
4. Always know more than you are expected to know.
5. Remember that difficulties are only more to overcome.

6. Never put your hand out further than you can draw it back.
7. At times bold, always patient.
8. Men say "What do they say?" Let them say.
9. Make good use of other men's minds.
10. Listen well, answer cautiously, decide promptly.
11. Preserve by all means in your power a sound mind in a sound body.

He had been patient, now it was time to be bold. The board agreed to move forward with the cannery. In January 1894 he jotted down an estimate of what it would cost to build:

Cannery building only:	$8,000
Machinery:	
boiler	1,000
washer	525
150 coolers @ $4	600
knife & hoist	400
other machinery	1,700
	4,325
Other buildings:	
mess & dwelling	1,200
store & office	700
Mngrs. house	800
China house	800
Indian shacks	600
	4,100
Clearing:	
5 acres @ $200	1,000
Boats:	
60 boats @ $45	2,700
Contingencies	2,575
	$22,700

THE GOOD HOPE CANNERY

Bell-Irving hired John Elliot as his construction foreman at $5 a day. Elliot's crew would include Archie McNeal, George Laforce, W.A. Burris, Malcolm Immet, Leslie McNeely, Stanley Mitchell and a cook at $25 a month. The lumber—90-foot-long (27-metre) timbers of knot-free, No. 1 Douglas fir—would come from Victoria Cannery's sawmill, located on the Wannock River opposite Rivers Inlet Cannery. Cedar shakes were purchased from Dan Bella Bella, presumably from Bella Bella. Construction got under way in early April, the carpenters using mortise and tenon joinery with dowels and square nails.

At what point did Bell-Irving decide on a name for the new cannery? Canneries in Rivers Inlet tended to be named after their owners (Wadham's, McTavish), a physical feature (Wannock, Goose Bay), or they were given an unmistakably British handle (Strathcona, Victoria). Good Hope is the odd name out. If it wasn't Henry Bell-Irving who named it, it was certainly he who approved it. He had a couple of precedents. South Africa's Cape of Good Hope was named in the fifteenth century by Portugal's King John II because of great optimism surrounding the opening of this sea route to India and the East. In 1805 the Northwest Company established an outpost in the lower Mackenzie Valley, naming it Fort Good Hope, no doubt for similar reasons to John II. Bell-Irving and the ABC Company must likewise have been optimistic about their prospects. Another Good Hope, the British Royal Navy's battle cruiser, HMS *Good Hope,* had its hopes dashed in 1914 when it was sunk at the cost of nine hundred lives by German Admiral von Spee in the Battle of Coronel.

Whoever named the cannery, its first manager would be C.A. Sutter. Carl Anton Sutter was born in Sweden around 1869. He arrived in Texas at the age of ten, later moving to Astoria, Oregon, to work for his Uncle Holm, a Chinese labour contractor. Chinese cannery workers referred to Sutter as "the King" and his wife, naturally, as "the Queen." The nicknames stuck to the extent that even their grandchildren referred to them as "the King and Queen."

"He had a phenomenal memory," grandson Carl Sutter told me in August 2009. "The King was called to testify at a fisheries commission hearing in Washington, DC. As he was speaking, citing facts and figures, he was asked if he'd like to refer to notes for the sake of accuracy. He waved it off, telling them he didn't need to."

Sutter married Azubah Decrow Fain and they raised three sons. He managed Good Hope for two seasons before becoming the general manager of ABC's subsidiary canning operation, Fidalgo Island Packing Company, in 1897. Headquartered in Anacortes, Washington, Fidalgo grew to include canneries in Alaska. The King earned a princely annual salary of $7,000 at a time when the average American family income was about $750 a year and Harvard University professors earned about $4,000. In 1908 Sutter and Bell-Irving, according to a report in the *Anacortes American* newspaper, "joined in an enterprise which will add another considerable salmon cannery to the six establishments in Anacortes proper."

Grandson Carl continued, "The King ended his career at sixty years old because he was just plain tired of working. [About 1931, the same year Bell-Irving died.] He'd spend his retirement down along the Seattle waterfront in his long black coat and black bowler hat whittling on a piece of wood. People referred to him as

Henry Bell-Irving and his wife, Ysabelle. Torquay, England, 1885. Photo courtesy of Ian Bell-Irving. Photo courtesy Patricia Wilson.

THE GOOD HOPE CANNERY

'Old Man Sutter.' If the King was whittling on a piece of wood, it was not the time to talk to him." He died in 1954.

But fifty-nine years earlier, on May 4, 1895, he was not whittling wood on a dock in Seattle; he was at Good Hope greeting H.O. Bell-Irving who, after a 29½-hour trip from Vancouver aboard the Canadian Pacific Navigation Company's *Danube*, had just disembarked. Bell-Irving saw with satisfaction that most of the cannery's foundations were finished. He then climbed the hill south the cannery and took a number of photos. In his journal he

Good Hope cost $30,791.70 to build and equip, including 73 skiffs—a small price to pay for the large returns investors would enjoy. This photo was taken by H.O. Bell-Irving in 1895.

MEN OF GOOD HOPE

recorded the timing for the arrival in a few weeks of everything from oarlocks and sails to steam gauges and acid jars.

Bell-Irving reboarded the *Danube* and toured up and around the inlet, dropping in on H.J. Kirkland at Wannock Cannery and on G.S. McTavish at the Rivers Inlet Cannery. McTavish managed the cannery from 1892 to 1897 and again from 1899 to 1913. In 1918 he built the cannery named after himself, selling it in 1920 to Gosse-Millerd Packing Company. The cannery ended up in the hands of BC Packers and in 1932 it was sold to ABC, which closed it forever in 1939.

On May 7 Bell-Irving visited a parcel of land that ABC owned in Schooner Passage, noting

Bell-Irving would much rather have been thought of as a big game hunter than a businessman. Shown are some of his trophies at his home on Seaton Street, Vancouver. c.1899. Photo courtesy of Patricia Wilson.

that the site—about a mile and a half down the passage—was a good spot for building with a plentiful supply of water. That afternoon Kirkland's steam tug was late in picking him up, so Bell-Irving, Ben Legeuse and someone identified only as "an Indian," rowed to Wannock Cannery at the head of the inlet, arriving at 7:30 p.m. Shortly after, Kirkland and his steam tug arrived, having called at Good Hope and discovered Bell-Irving and party already gone.

Business attended to, it was time to hunt. The next morning Bell-Irving, "Lazy Louis" and "Tinshop George"—an Oowekeeno chief—left in a canoe bound for the mountains surrounding Owikeno Lake. One thousand feet (300 metres) up, as the snow flew, Bell-Irving shot a big ram. The next morning the party canoed to the other side of the lake and spotted a goat on a ledge. They climbed and Bell-Irving shot a small yearling ewe, the animal tumbling 300 feet (90 metres). After lunch, they headed back to Kirkland's cannery for the night.

The next morning the hunting party resumed. They followed a stream for an hour up

THE GOOD HOPE CANNERY

a mountain, finally coming out of heavy rainfall. Bell-Irving had a goat in his sights, but in some way that he doesn't explain, his companions "made a mess of it & spoilt my chance, which was a sure one." He fired at the disappearing animal, but it was a long shot and he did not hit it. He was not pleased and noted: "Another time depend on yourself after goats—let Siwash [Natives] find them only." By 7:30 that evening he was back at Good Hope where the main cannery building rafters were all up. The next morning he boarded the *Danube* bound for Victoria, but not before taking photos of Good Hope from various vantage points.

Fishermen's cabins and bunk houses are shown on the right. The building in the centre with the ramp was used for net drying. c.1905–1915. Photo courtesy of Ian Bell-Irving.

Bell-Irving returned to Good Hope at the beginning of the third week of June. He "found all well" with seven Indian houses built and one on the way, two "white men" shacks completed, the store enlarged to 16 by 48 feet (5 x 14.5 metres), 9,300 new cases of cans made, the boiler fixed, two steam boxes made and all three retorts in position with "steam on." At last, the cannery was up and running. Out on the water, four of five scows were built and there were forty-nine boats on hand. He noted that he liked the "Jap boats best of all."

All the fishermen were lined up, but Sutter was not happy with his Chinese foreman, who

Good Hope's prime residential real estate was south of the cannery, a good location for the manager's house, c. 1900. Photo courtesy of Ian Bell-Irving.

THE COLLECTOR

In March 2008 I caught a ferry to Nanaimo and drove north to the town of Sayward to visit Robert and Nancy Critchley. Tony Allard had told me, "Rob is an extraordinary guy. He's an expert on everything connected to canneries."

The Critchleys' Jack Russell terriers greeted me when I arrived, and Rob and Nancy welcomed me into their home. Over tea and fresh-baked muffins we talked about fishing, canneries and Rob's passion for all things connected to the fishing industry. Rob and Nancy were once commercial fishermen. "Nancy was the better fisherman," said Rob and she didn't disagree with him. But by the 1990s fishing had proven too difficult for them to make a living, so they quit it. Nancy worked for a number of years for the Department of Fisheries and Rob now worked as a faller in the logging industry.

"Would you like to see the cannery?" asked Rob.

When he learned in 2003 that they were planning to demolish the Alert Bay Cannery, Rob bought the front section of it, painstakingly catalogued every plank of it, loaded it onto a barge, towed it to Sayward and, with the help of his father and a few others, meticulously reconstructed it on his property. He put everything back together just as he'd found it: light fixtures, signs, even the 1940s' caricature of Adolf Hitler a worker had drawn on a wall. But this was no empty shell of a cannery. Rob had installed a vintage canning line, including an "Iron Chink."

The Iron Chink was a salmon-butchering machine invented in Seattle in the early 1900s by Canadian E.A. Smith. A vertical wheel carried salmon past razor-sharp knives and cleaning attachments at the rate of one fish per second, doing the work of thirty or more Chinese butchers in a fraction of the time. A good butcher could remove fins, head, tail and entrails with eight knife strokes and dress up to two thousand salmon in a ten-hour day, but he was no match for the Iron Chink and by 1909 more than sixty were in use—more than likely including one at Good Hope—replacing thousands of Chinese salmon butchers.

"It's a complete canning line," said Rob. "It's not operational, but it's all in place."

We started at the end of the building where a fish ladder would have brought the fish into the cannery and, as we followed the canning line, Rob explained what each piece of equipment did. Some of the machinery looked long past ever running again but he assured me that all it takes is time. "They built these things to last in those days."

Looking at the rust and grime, it seemed impossible to me, but I didn't doubt Rob for a second. We came to three or four tables that looked new. "I haven't been able to find originals of any of these tables, these are the ones they'd gut the fish at, so I made these replicas." The excellent workmanship was evident even to my untrained eye and Rob told me even the paint colours were authentic.

We went upstairs to the second level. It was full of pallets of old nets, cork floats, copper sulphate (bluestone), old VHF radios, radar equipment, outboard engines, flags, lights, oars and oarlocks and a hundred other items, large and small, mechanical and chemical, that were once part and parcel of the fishing industry. He showed me his collection of old Easthope and Vivian engines, the one- and two-cylinder "lungers" of the early days.

"This one," he said, pointing to an Easthope, "had a hole in the cylinder wall, so I patched it over with some liquid metal. It's stronger than ever."

It was astounding that he was able to take engines that had been discarded as junk forty and fifty years earlier, some recovered from the ocean floor, and not only clean, repair and restore them to working condition, but smooth, polish and paint them into mint condition. These engines—indeed everything Rob touched—lived again, the way a meticulously restored collectible car lives again. We returned to the ground floor and Rob showed me a 1930s' gillnet boat and an old fish packer boat that he was restoring.

"So when you're done," I observed, "you'll have a vintage working cannery, a vintage fishing boat, and a vintage boat to collect the fish. You'll have the whole thing."

"Yeah, well, that's the plan," he said with a grin.

needed to be replaced because he had no authority with the Chinese labourers. "The King" was in a position to know, having initially learned the business of recruiting and retaining Chinese workers from his Uncle Holm. The Chinese were treated abysmally by any measure. Bell-Irving told a royal commission looking into the issue of Chinese labour that they "won't strike when you have a big pile of fish on your dock. They are less trouble and less expense than whites. They are content with rough accommodation at the canneries… I look upon them as steam engines or any other machine, the introduction of which deprives men of some particular employment but in the long run, it enormously increases employment." The "rough accommodation" of which he spoke was the "China House." At Good Hope it stood on the rise behind the south end of the cannery, easily the least desirable location. A two-storey structure, it housed Chinese workers in the most cramped manner imaginable. Three-tiered bunks provided little or no privacy. Beds were narrow and wooden, and no mattresses were provided. A small dining hall was heated by a wood stove. Chickens and pigs were penned outside nearby. Cooking was done under an outside lean-to.

While Sutter dealt with the issue of the Chinese contractor, assistant manager Woods dealt with a missing package of tobacco.

Rivers Inlet 23rd May, 1895

I am ordered by the manager Mr. Sutter to advise you that a shortage of one package of tobacco occurred in the cargo of the S.S. *Coquitlam* invoiced by Mssrs. Oppenheimer Bros. … The remainder of this invoice arrived, but this package did not appear on the ship's manifest, and as there was no shipping receipt it is hard to locate the error. In the meantime the invoice has now been certified to.
I await your instructions.
I beg to remain

Yours very truly,
R.J. Woods

Meanwhile, no detail escaped Bell-Irving's attention either. He wanted the piles braced to the wharf and ordered that no new buildings were to be erected between the Indian shacks and the cannery or between the mess house and cannery. Also, the flume from the corner of the knoll was to be raised high enough that it would deliver water into a tank that stood eight feet (almost three metres) above the cannery roof. Finally, he noted with annoyance that Good Hope had received no fruit except for three boxes of oranges and that his supplier had been sending poor stuff and overcharging for it to boot.

Sutter had been at Bella Bella since June 20 recruiting Natives for Good Hope. There had been some sort of "a fuss" in securing them, but he had prevailed. The Oowekeeno were the first Rivers Inlet fishermen, using traps, hooks, spears and, even before Europeans arrived, fibre nets to catch salmon. The families of the Native fishermen were as important as the fishermen themselves. "The desirability of a particular Indian," said a report to the BC government, "is measured by the number of women his household will produce for the canneries as fish cleaners and can fillers." Two years later Sutter would contract with Natives at Port Chester and on the Nass River:

> We the undersigned Port Chester Indians agree and bind ourselves to fish for the Good Hope Cannery on Rivers Inlet during the Sockeye Salmon season of 1897 until such time as the Manager of the said Good Hope Cannery shall order us to stop.
>
> Also that all fish shall be delivered into the Company's scows each day.
>
> In consideration for the above mentioned Sockeye Salmon the sum of 6¢ (six cents) for each sockeye will be paid, final settlement to be made at the end of the fishing season. And also a free passage will be furnished from Port Simpson to the Good Hope Cannery to all captains and pullboats by the said Good Hope Cannery.
>
> No women allowed to pull boats or accompany fishermen to the camps.

Canning required manpower—lots of it. Attracting workers to the canneries was seen as everyone's problem, and so the canneries worked co-operatively to attract as many as possible. But poaching fishermen wasn't exactly unheard of. To prevent it, contracts were drawn up.

Good Hope Cannery Managers

C.A. Sutter	1895–1897
R.J. Woods	1898–?
R.E. Carter	?–1920
Victor Larson	1921–1935
Levi Lauritsen	1936–1950
Oly Anderson	1951–1969

Of course, they were often only as good as the paper they were written on and various kinds of bonuses—cash or gear—were sometimes offered as inducements to break contracts and switch allegiances.

By 1900 there were six canneries in Rivers Inlet packing a total of 75,413 cases annually: Rivers Inlet Cannery, Wannock, Good Hope, Brunswick, Wadham's and Vancouver. The number grew, peaking in 1925 with eleven packing a total of 197,087 cases. As quickly as the canneries multiplied, the Oowekeeno population declined. Disease, firearms and alcohol took their toll. In earlier times, in the fur-trading period, they had at least maintained their independence. But the rapidly expanding commercial salmon fishery was not built on trading, but on labour. Hungry for fishermen and shore workers, the canneries recruited from tribes as far away as the west coast of Vancouver Island. These workers were augmented by Chinese and Japanese immigrants.

At the turn of the century most of the fishermen working in Rivers Inlet were Natives, approximately a quarter were Japanese, and the rest were Europeans: Norwegians who had settled in the Bella Coola Valley in the 1890s and Danes who had colonized Cape Scott on Vancouver Island. The final wave arrived in the early 1900s when Finns established the community of Sointula on Malcolm Island. With a relatively small population to begin with, in sharp decline there simply weren't enough able-bodied Oowekeeno workers available to catch and process the fish for the explosion of canneries, and those who were able were already engaged. Canneries could always replace the Native fishermen with the more productive Japanese who delivered twice as many sockeyes as Europeans and nearly three times as many as Natives, but the families of Native Fisherman working in the canneries were irreplaceable. To find Native fishermen and cannery workers, Good Hope had to look elsewhere.

MEN OF GOOD HOPE

Bell-Irving welcomed Sutter and Kirkland aboard the *Danube* on June 23. Two days earlier the *Fingal* had arrived at Good Hope from ABC's Phoenix Cannery in Steveston with a cargo that included 3,193 cases of cans, 10 tons of coal and 14 boats. Returning to Good Hope, Bell-Irving and companions passed the steam tug *George* coming from Kitimat with a long string of canoes in tow. The next morning, in the pre-dawn darkness, Bell-Irving and Sutter got into a boat and rowed from the *Danube* to Good Hope.

Jotting down the actual built size of the cannery—343 by 76 feet (104 x 23 metres)—Bell-Irving tallied the cost so far to build and equip Good Hope. Including machinery and tools, boats and scows, the mess house, construction and other expenses, the total stood at $18,953.67. (According to Ian Bell-Irving, the value of the 1895 pack was enough to pay for half the cost of building the cannery.) All in all, it had been a satisfying trip. He had a ram and a ewe head for trophies and he had a functional cannery.

Sutter arrived in Vancouver in mid-September with Good Hope's statement of accounts. The cost of the pack of 19,038 cases, excluding scows and boats and freight down to Vancouver,

Henry Bell-Irving kept meticulous records, mostly business related, in identical, leather-bound, breast pocket–sized notebooks. This page, from 1896, records the costs of constructing and equipping Good Hope. Image courtesy of the City of Vancouver Archives.

THE GOOD HOPE CANNERY

The families of Native fishermen were as important to the cannery operation as the fishermen themselves. City of Vancouver Archives: 173-B-7 album M-3-30.1.

came to $2.49¼ per case. On their biggest day, they had cooked 1,400 cases. Of the canned salmon at Good Hope still awaiting shipment, there were 550 cases of Holly brand, 1,500 of Lynx, 503 of Doouen and 7,835 cases that were unlabelled.

Before returning to his home in Seattle, Sutter discussed the next year's season with Bell-Irving. Sutter was impressed with the boats made by Japanese builder Kuno and recommended more of them be purchased at $35 each. The conversation turned quickly to

MEN OF GOOD HOPE

These shiny, silvery cans, ready for filling in the loft at Good Hope Cannery, eerily resemble the skin of the salmon they are about to contain, c. 1895. Photo courtesy of Ian Bell-Irving.

Sutter's plans. Was he, asked Bell-Irving, amenable to managing Good Hope next season? He was, but at a salary of $1,400. Bell-Irving agreed to it at once. That settled, Sutter listed the key men he wanted: Ben Legeuse for watchman and fisherman boss; Humphries for net man; and George Laforce for engineer and blacksmith. On the operations side of the equation, he wanted an extension of the fish house, more room for nets, one new soldering machine and 35 new nets, 40-mesh deep with lines to match, to provide for 110 boats in total. Done.

THE GOOD HOPE CANNERY

With Sutter gone, Woods, the assistant manager, held the fort at Good Hope. On October 3, he wrote to head office:

Mssrs. H. Bell-Irving & Co.
Vancouver

Gentlemen
Since my other enclosure was written Capt. Myers decided to take only 1000 cases instead of 2000, saying that he was to take 2000 cases "if the *Boscowitz* had not done so," however he appears to have no room for the other thousand. I have accordingly changed shipping receipts for the 1000.

Yours truly,
R.J. Woods

Woods departed not long after, leaving Ben Legeuse to look after the cannery. For $50 a month minus board, Legeuse would spend the winter fixing s and chopping the cordwood that was piled on the wharf, all one hundred cords of it. As the long winter set in, those remaining at Good Hope were not about to be forgotten, writing this complaint to head office:

7th November, 1896
H. Bell-Irving & Co.
Vancouver

Dear Sirs,
Last Wednesday evening we sent a boat out to meet the *Boscowitz* to put our mail on board, as she had not called in here, and although we were within 100 yards of her, she would not stop and take the bag on board. It's now 3 weeks since we had a mail here, please call someone's attention to this treatment.

We beg to remain,
Your Obedient Servants,
Good Hope Cannery

MEN OF GOOD HOPE

From 1895 to 1910 Good Hope was serviced by the Boscowitz Steamship Company's iron-hulled *Venture* and its *Barbara Boscowitz*, a 396-ton ship described by J. Turner-Turner in his *Three Years Hunting and Trapping in America and the Great North-West*:

> We all embarked on the *Boscowitz*, the most miserable, dirty little steamer imaginable... She starts at no regular date, and takes her own time, her average speed being five or six knots an hour... the cabin swarmed with cockroaches, which are repulsive enough when occasionally found in the kitchen, but almost unbearable when as bed-fellows... At breakfast the food was so badly cooked and appeared so uninviting, that it was with difficulty we could induce one another to partake of anything... The voyage we found tedious and monotonous... owing to the slow pace and delays... our vessel was crowded; the saloon with prospectors and adventurers... The remainder of the boat was filled with all sorts and conditions of men and women, including several Indians, one of whom was dead...

Steam, steel and sweat: canning at Good Hope. Photo courtesy of Ian Bell-Irving.

Chapter 3

The Letter Book

The [eighteen] nineties witnessed the sunset of the Victorian ethic, the passing of a time when the role and importance of God, the Queen, the flag, duty, honour, virtue and family life were all clearly defined. It was the twilight of an age, the end of a century, when one good opinion, on any topic, could last a lifetime; when there was right and wrong without shades in between; when there was no confusion about what to believe and what to trust. Faith in the church, the British Empire and hard work gave shape and stability to life.
—June Callwood, *The Naughty Nineties*

The University of British Columbia Archives houses eighty-seven boxes of ABC Packing Company records. In Box 52 is a one-of-a-kind book. Buried among accounting ledgers, the "Letter Book" contains copies of dry goods and merchandise orders, fishing contracts, damage reports and other outward-bound business correspondence from Good Hope Cannery for the years 1895 to 1899. In those pre-carbon-paper days, a letter would be written in ink on company stationery and before it dried it would be inserted between pages in the Letter Book. The tissue-paper-thin page would absorb the excess ink and a copy of the letter would be created.

The book itself is in superb condition, its spine straight and strong and its pages intact, if a little yellowed with age. The ink is light to dark brown, a little blurry from the blotting and smudged in places, but overall it is easily readable. Turning the pages you are transported back to Good Hope as it was in the 1890s. You are at a desk in the office in the building to

THE GOOD HOPE CANNERY

the southwest of the cannery, the one that eventually came to be known as "the Hilton" and which was demolished in 2006, the one where the seven accounting files were hidden in a wall for sixty years. You are standing at the elbow of Mr. R.J. Woods, bookkeeper and assistant manager of the Good Hope Cannery; it is July 22, 1895, the middle of the cannery's first season of operation. Woods is penning an important order. Is it for more tin plate? More skiffs? What is so desperately in short supply? You lean in as he orders… six black cashmere shawls, two dozen fine silk handkerchiefs, two dozen men's yellow-coloured half hose, and one dozen white Turkish towels. Fine silk handkerchiefs and yellow socks? Turkish towels? Are they canning fish or trying to be Bloomingdale's?

Next, he writes to Mr. John Neugel, watchmaker, of Victoria, requesting that he repair the gold pocket watch that he's sending via the steamship *Danube*. Well, accurate time is important, of course, but where is the stream-driven cacophony of the cannery? Where are the tons of sleek, silver-scaled sockeye, the steaming retorts, and the fish-bloody, flashing knives of the skilled Chinese? "We are now returning you the two lines of mens shoes 957B and 696B as they are exactly what we have been trying to avoid buying, as we are overstocked with them," he writes tersely to the Ames Holden Company, one of Canada's largest shoe manufacturers. Men's shoes, is this any way to run a cannery?

Skipping forward in the book to the 1896 season, there's an order from manager C.A. Sutter himself. Please send "one set good Boxing Gloves." Boxing gloves? Surely he means gloves for handling boxes? Sutter fades away and once again you're standing beside Woods as he prepares an order for laundry soap, Mikado brand, named for the Gilbert and Sullivan comic opera. Oh, and please send us "Hungarian" flour, milled from hard wheat grown at higher altitudes, also known as winter wheat. Nothing else will do. But wait, also, 50 boxes of "Pilot Bread." Pilot bread is more commonly known as hardtack, the hard, dry, flour-based biscuits that were an armed forces staple for hundreds of years. Now that's more like it: tough, spartan, tasteless food. Hold on, what's this? Two dozen good men's straw hats and two dozen briar pipes? And you thought cannery life was no picnic. What's this? Woods is penning another letter. It's the end of the season and he's listing the brands and the number of cases still in the cannery:

Holly Leaf, Lynx, Salmon Fly and Red Star. So it really was a cannery after all.

Over and above the information itself is the archaic business-letter form, the salutations and closings, the polite and mannerly style of the language. Commonplaces of the era, but endearing to read now. It wasn't just *yours truly*, it was *believe me yours truly*. Employees were *your obedient servants* and they would *beg to remain* so. The recipient of a letter *will much oblige* the sender by replying promptly.

But the most colourful finds in the book are the letters sent by an incensed R.J. Woods. While we only have Woods' side of the story, it seems that a Reverend S.S. Osterhonk of Nass River told a number of cannery men in May 1897 that Woods had offered 7¢ a sockeye to the Nass River Indians, a penny more than the industry's agreed upon rate. Among the cannery men were N.H. Dempster, John Draney and Robert Cunningham. Cunningham had established the village of Port Essington at the confluence of the Skeena and Ecstall rivers in 1872 and built a cannery there: Cunningham's or the Skeena Cannery, as it was also known. John Draney was a member of the well-known Draney cannery clan, after whom Draney Inlet is named. Dempster has proven elusive. Captain Myers of the steamship *Danube* was also implicated. The irate Woods, his pen dripping with sarcasm, would not be maligned and goes on the offensive.

> 19th May, 1897
> Rev. S.S. Osterhonk
> Nass River
>
> Sir,
> I learn from Mssrs. N.H. Dempster and R. Cunningham of Skeena that you informed them of having a letter written by us offering the Nass Indians 7¢ per fish for this season at Good Hope Cannery. Would you oblige us by sending us a copy of same, or in case said letter be mislaid, mention to whom said offer was made, by so doing you will much oblige.
>
> R.J. Woods
> 10 June, 1897

THE GOOD HOPE CANNERY

Mr. J. Draney
Nass River

Dear Sir,
I am informed that you told Capt. Meyer of the "Danube" that I had offered the Nass fishermen 7¢ for sockeye for this season. You without doubt read my letter or you would not make such a statement so will you please advise us to whom the letter was addressed or the above price quoted, in writing of course. By replying at your earliest convenience, you will much oblige.

R.J. Woods

24th June, 1897
Rev. S.S. Osterhonk
Nass River

Dear Sir,
While thanking you for your favor of 21st June replying to my several letters, the question as to "name of your informant" remains unanswered. I made mention that R. Cunningham of Essington has quoted you as his authority for this and as you are not supposed to be in the lying and slandering business, it might be well on your part to contradict him unless he is assisting you to do so, kindly answer the following questions.

Have you ever seen a proposal from here for J. Draney fishermen for Good Hope Cannery?
What is the name of the person who told you I offered more than 6¢…?
Did you ever tell Dempster and Cunningham or either of these that I had offered 7¢ season for fish?

Much regretting being obliged to cause you this trouble and requesting a reply at your earliest convenience.

Believe me yours truly,
R.J. Woods

24th June, 1897
Draney

Dear Sir,
On the 10th last I took the liberty of writing you as to a statement you made to Capt. Meyer regarding an offer made by one of 7¢ for salmon at this cannery. You have perhaps not received my letter so you will much oblige by mentioning the name of your informant.

  Your kind and prompt attention to this will much oblige.

Yours truly,
R.J. Woods

July 15, 1897
Draney

Your determined non-acknowledgment of my second letter will convince everyone that you are not only a liar, but also as ignorant and ill-bred as you are untruthful. I might suggest that you and the Rev. Osterhonk cast lots to decide which is the most deceitful.

  I am taking all pains and successfully in bringing the honest manly qualities of both of you to the notice of as many persons as possible. Not even knowing you by sight I may not have an opportunity of expressing my idea of such a unique liar to you personally. Trusting you will read this to Capt. Meyer of the *Danube*.

Believe me yours truthfully,
R.J. Woods

  In May 1896 Henry Bell-Irving travelled to Rivers Inlet on another grizzly bear hunting expedition. Upon arrival, Owekeeno chief "Tinshop George" told him that he had recently killed a "misatchee bear"—a wicked bear. His appetite whetted, Bell-Irving and party explored Loquaish Creek and the Dallig, Nuhantz and Shoemahout rivers in a spoon canoe. Spoon canoes, with their scoop-shaped sterns and bows resembling handles, were hand built from

THE GOOD HOPE CANNERY

Natives at Rivers Inlet (possibly Wadham's Cannery) boarding the S.S. Boscowitz at the end of the fishing season sometime in the 1890s. City of Vancouver Archives: 173-B-7 album M-3-30.2.

logs by coastal Natives for use in river transportation and eulachon fishing. Unfortunately, Bell-Irving had no luck and by June 1 was empty-handed at the mouth of the Wannock, awaiting the *Danube*.

He was much luckier with the fishing season. Sutter told him that the salmon run had been so heavy that one day they received 43,000 fish. The pack had come in at 30,100 talls, cylindrical cans holding one pound (.453 kilograms) of salmon. The cannery had employed 116 Chinese and about 150 "klootchmen," a Chinook slang word for Indian women. Sutter was happy with the performance of their Chinese foreman, H.G. Tay of Victoria, a "good man" who, with the backing of Chu Chung & Co., would return, they hoped, for 1897. Ben Legeuse was staying on over the winter as watchman. Sutter recommended the purchase of a new top-tightening machine, estimating it would eliminate the need for as many Chinese "wipers, up to 18 hands," and would save the business somewhere between $500 and $600 a season. Wipers cleaned cans with old linen netting rags before Chinese "toppers" placed a piece of tin on the salmon pieces in the cans and positioned the lids for tightening prior to cleaning in hydrochloric acid and dissolved zinc, in order to ensure strong bonding when the tops were soldered.

In May 1897 Bell-Irving arrived at Good Hope and observed the cannery in operation: "Cleaning and washing all done in gut shed before fish sliced by circular knives. [Gang knives: knife blades attached to a shaft with a lever at one end turned manually.] Slicing done in cannery and pieces taken direct to filling tables. 24 Chinamen and 24 klootchmen in gut shed… 50 fillers, 20 toppers, 8 test kettle and 12 in bathroom. When making cans, 13 seamers and 8 toppers. One man cutting tops, cuts 5 boxes tin per day."

But there was more to Bell-Irving than a head for numbers and production. Not only was he a canner, he was a canny marketer. Up to the time the ABC was formed, salmon canners had generally played up the river of origin in marketing their product, such as British Columbia Canning Company's O-Wee-Kay-No River brand or J.S. Deas' Fraser River Salmon. But early on Bell-Irving emphasized "sockeye" in his marketing efforts. He was so successful with this strategy that "sockeye" became synonymous with the highest-quality salmon.

Canning labels from the Anglo-British Columbia Packing Company. Top: One of the earliest ABC brands. Bottom: Bell-Irving's heritage reflected in ABC Co's "pink" brand. City of Vancouver Archives.

On May 8, 1909, 53-year old Henry Bell-Irving and a party that included his son Dick boarded the *Beatrice* and departed Coal Harbour bound for the Kemano River where he planned to hunt grizzly bears. Their route took them to Powell River, Alert Bay, Good Hope, Namu, Swanson Bay and Hartley Bay.

1911 Census

In 1911 the Canadian government conducted a census. It reveals that fifty-three men, women, and children of Caucasian or Asian descent were resident at Good Hope on June 1. They were of Finnish, Danish, German, Swedish, Norwegian, French, Scottish, Australian, Irish, Chinese and Japanese origin. Six religious affiliations were present: Lutheran, Methodist, Anglican, Presbyterian, Roman Catholic, Confucian and Buddhist. There were three unmarried, Canadian-born, Caucasian fishermen from Ontario: 43-year-old William R. Camaron; 50-year-old B.S. Glassford; and 46-year-old Henry N. Dun, a French-speaking Catholic. The only other Canadian-born Caucasian at Good Hope that year was foreman James Harris, a 44-year-old married Methodist from New Brunswick. The bookkeeper was Quentin R. Young, a single, 25-year-old, Scottish-born Presbyterian who had immigrated to Canada the previous year. Walter Russell looked after the nets. Like his countryman Young, the unmarried 28-year-old Russell had immigrated in 1910. Were Young and Russell friends from the old country? Did one introduce the other to work at Good Hope? Keeping the cannery running was engineer N.J. McIntyre, a 29-year-old Australian who had immigrated in 1908.

In 1911 approximately seventy-six skiffs were in use at Good Hope. Approximately fifty-seven were manned by Native fishermen. Photo 1930, courtesy of the United Church of Canada.

> In what must be an interesting story (alas, unrecorded), an 86-year-old Irish Anglican bachelor named William Pattinson kept the store.
>
> The labour contractor was T. Nakato. He was joined by his wife, three-year-old son and six-month-old daughter. Other Japanese kept company with the Nakatos. Y. Nici was a 40-year-old fisherman joined by his sons, 16-year-old Y. Nici junior and 12-year-old K. Nici. M. Shinde, I. Nagatani and Y. Naganethi rounded out the contingent of Japanese fishermen. Twenty-three Chinese men worked inside the cannery. They were mostly single, entirely Confucian, and they ranged in age from 19-year-old Dat Lee to 60-year-old Jung Yim. Good Hope's kitchen was manned by Chung Lee Long, a 27-year-old who had immigrated to Canada four years earlier.
>
> Good Hope's Native fishermen and female cannery workers are not covered in this census because on June 1 they would have still been in their villages and, at any rate, at that time Native peoples were accounted for in a separate census. It is possible to estimate, however, that with approximately seventy-six skiffs available (using the number from the 1903–4 boat-rating system) and only thirteen Caucasian and six Japanese fishermen present, the other fifty-seven boats had to have been manned by Native fishermen or by other fishermen who arrived later in the season.

After picking up Indian hunting guides Paul Charlie and Johnny Paul, they made their way to the mouth of the Kemano River. On May 15, Bell-Irving and the two guides started up the river, having arranged to meet the *Beatrice* in five days' time. Once again, Bell-Irving was to be denied the prize he coveted. They spotted a large grizzly on a snow slide as they were returning to camp one evening, but hunting him in heavy rain the next morning "he got into thick bush on the mountain and [we] could not follow," noted Bell-Irving with disappoint-

that afternoon they met up with the *Beatrice*, just one black bear to show for their fifteen-mile, five-day trek.

Back at Kemano village on board the *Beatrice*, Bell-Irving caught a 33-pound (15-kilogram) spring salmon (or chinook) and one of his guests a 45-pounder (20.5 kilograms). The flesh of the two fish was "pale pink, nearly white," noted Bell-Irving. The colour is due to the salmon feeding on herring, sardine and other white-fleshed fish, as opposed to the shrimp eaten by red springs. The taste of a white spring is much oilier and more flavourful than a red.

Bell-Irving (second from the right), always a commanding presence, strikes a confident pose in the company of Native hunting guides in Lillooet, about 1919. Photo courtesy of Patricia Wilson.

The next day they were in a small bay off McAlister Point trolling for salmon, spearing crabs and scanning the snow slides for bears. Spotting one, Bell-Irving, ignoring the dangerously unstable slopes, went ashore. He was getting into position to shoot when a snow slide came straight at him. He began to run as if his life depended on it, which it did. "Fastest sprinting for many years," he noted.

On their way back to Good Hope they stopped in at ABC's North Pacific cannery on the Skeena River, where new buildings had gone up including forty Indian homes.

It was 10 p.m. and dark when they came through what they supposed was Schooner Pass into what looked like Rivers Inlet. Bell-Irving couldn't find anchorage, so he lay to. "Dead calm," he noted. If not lost, they were at least unsure about their location. "In an inner bay in labyrinth of islands I watched till 2:30 a.m. when found that anchor was holding, so turned in," wrote a concerned Bell-Irving. He was up two hours later and they started off. The *Beatrice* was now dangerously low on gas, just two inches in the starboard and port tanks, but recognizing Safety Cove and Cape Calvert they at least knew where they were. They tied up at Good Hope at 7:00 a.m., greatly relieved, and took on 93 gallons (423 litres) of gas. Just over an hour later they were gone again and on June 4 they tied up in Coal Harbour, the trip over.

Fishing in Rivers Inlet was 24/5, Monday to Friday, on board your skiff. Here an ABC Co. packer boat (left) is getting ready to tow skiffs to the fishing grounds. Photo courtesy of Patricia Wilson.

Chapter 4

Catch What You Can, Can What You Catch

*Now some of us think of the future
While others have things to forget,
But most of us sit here and think of a school
Of sockeye hitting the net.*

From "The Song of the Sockeye" by Ross Cumbers, c. 1940. The song was found under a glass-covered notice board at Wadham's Cannery, Rivers Inlet.

An album of photos that had once belonged to William Robert Jamieson was given to the City of Vancouver in 1938. Jamieson's album identified in white ink on the album's black pages people, places and events in Vancouver from around 1900. Some of the earliest photos are of the aftermath of the 1898 fire that destroyed much of downtown New Westminster. There are also images of Seymour Creek, a Port Moody logging camp, Stanley Park, the SS *Danube* leaving Vancouver bound for Skagway and the Klondike gold rush, and of Vancouver streets on September 28, 1901, when the Duke and Duchess of Cornwall and York paid a visit. There are two pictures of a "machinist's picnic" and photos of Balmoral Cannery on the Skeena River. There are also a number of photos dated "189–" depicting Rivers Inlet canneries and domestic cannery life including a few taken in and around a cabin identified as being at Good Hope. In one, Jamieson is shown washing his laundry. In another, he is posing in the salal in front of the same cabin with a group of unidentified men. And in another, he is seated on a bed tying his shoe. Above him hangs the Union Jack. To his left on the wall are a dozen photos suspended in a fishing net. Two lines from Gary Geddes' 1971 book of poems, *Rivers Inlet*, came to mind: "The nets of memory. Black caves extend from all our lives like ellipsis." What was my research but a giant net, cast to collect memories?

Cannery millwright Robert Jamieson, living on the outskirts of the British Empire in Rivers Inlet, 1899. Note the Union Jack and old fishing net with its "catch" of photos. City of Vancouver Archives: 173-B-7 album M-3-29.5.

Not every photo is of Jamieson and his European cohorts. There is a formal-looking picture of a young Chinese man posed beside a leather armchair. His right arm is over the back of the chair. In his left he holds a book at chest height, as if he were about to read from it. Behind him are delicate lace curtains. His name is not recorded, just that he was a cook at a Skeena River cannery, probably Balmoral. Why did Jamieson take his picture? What is the significance of the book the man is holding?

On the basis of the "189–" photos of Jamieson at Good Hope, he must have worked there sometime between 1895 and 1899. I hoped that I would encounter him elsewhere in my research, but after a year I was still in the dark. I returned to the archives and asked what information existed on him, and was handed a folder containing a personal profile and work history that he had provided to the *Vancouver Daily Province* in 1937.

Jamieson was born in Plainville, Ontario, in 1867. He moved to BC in 1891 and eventually worked for the Vancouver Street Railway Company before taking his first cannery job as a millwright engineer in 1899 with Dawson & Buttimer at their Brunswick Cannery in Rivers Inlet. I now had a year for the "189–" photos: 1899. He was with Dawson & Buttimer until 1907, then BC Packers and Wallace Fisheries before joining ABC in 1915 through 1919.

Laundry day at the Good Hope Cannery. Robert Jamieson hanging his clothes. c. 1899. City of Vancouver Archives: 173-B-7 album M-3-28.2.

So Jamieson worked for ABC, just not in 1899 when he was with Dawson & Buttimer, unless while working for them he also did some work at Good Hope.

~

In 1902 Good Hope's can fillers filled 12,000 cases of sockeye (576,000 pounds/261,800 kilograms) and 382 cases of spring and fall fish. In 1905 the cannery packed 16,400 cases that, at 48 pounds (22 kilograms) of sockeye per case, tipped the scales at 787,200 pounds (357,800 kilograms). How many individual sockeye did this represent? It took about 13 sockeye to fill a 48-pound case, so roughly 213,200 fish were caught and processed that season—one season at one cannery in an inlet that in 1905 had six canneries in operation. Of these 16,400 cases, over 1,600 were, according to ABC employee Ma Toi, improperly sealed by the Chinese workers and had to be done over, due to leaks in the cans, requiring re-soldering. Consequently, the Chinese contractor, a man identified as June, was out $600. June was responsible for the work of the Chinese labourers. If they failed to can salmon to the standard set by the cannery, he had to bear the financial consequences, consequences he passed on to the workers. But the workers disagreed with manager R.J. Woods' $600 assessment, and so it wasn't June the workers blamed for their lost wages but Woods.

Ma Toi surfaces again in the summer of 1921, attacked in his cabin aboard the SS *Camosun*. An unnamed Chinese assailant armed with a stolen butcher knife had slashed away at the unfortunate Toi for unknown reasons, sending him to the hospital at Ocean Falls. The attacker was arrested. As for Toi, "there was little hope for recovery."

The Rivers Inlet catch in 1908 was down from 1907, not for lack of fish, but for overcast skies and rain. "In no other district of BC is the salmon catch so much affected by weather," observes Cicely Lyons in *Salmon: Our Heritage*. "Sunshine and warm days appear to attract the salmon to the surface within reach of gill-nets; while in cool, dull weather the fish prefer deeper water."

Victoria's *British Colonist* newspaper reported on June 16, 1908, "Yearly the cannery foremen come to the West Coast [of Vancouver Island] villages to recruit labor and the Indians are at once laden on the Steamer and carried to the canneries for the season... Good Hope, Brunswick, Rivers Inlet and Wadham's." Two years earlier W. Woolacott of Victoria sailed on the steamer *Queen City* up the west coast of the Island. At Clayoquot he went ashore and "engaged over 100 Indians for Good Hope cannery." Arriving back in Victoria on June 7, he arranged for the *Queen City* to make a return trip to pick up the 100 Indians from Clayoquot and another 200 from other villages destined for other canneries before proceeding to Rivers Inlet. On June 20, 1907, the SS *Tees* under the command of Captain Townsend left Victoria "with every available berth occupied and overflow passengers sleeping on and under the dining room table." The ship picked up 75 Indians at Nootka and Kyuquot bound for Good Hope. On one such trip in 1915 a ship carried Chief John Jumbo and his baby daughter, Christine, born in Nootka on January 9 of that year. Unfortunately the little girl never made it home. A tombstone near the site of the Rivers Inlet Hospital, in even earlier days the site of the Vancouver Cannery, records that she died at Good Hope Cannery on July 23, just seven months old. By 1919 the *Princess Maquinna* "went up as far as Rivers Inlet... taking 175 Indians north to the canneries there."

Chinese workers were also carried north on these ships. A reporter for the *British Colonist* referred to them as "human freight," lumping them in with "tiers of cases of tin plate for salmon tins, bricks, pipe, pig lead, groceries in big piles, flour and cannery machinery." But what really captured the reporter's attention were the alarm clocks. "The Chinaman scarcely ever travels without his alarm clock," he wrote. "And when the *Queen City* sailed for the canneries there were many more Waterbury nickeled alarm clocks than have been on that vessel for a long time, for huddled about on her decks were nearly a hundred Chinese cannery employees bound to the scene of their labors. Other than alarm clocks the Chinamen had sundry crates of chickens, pigs and other livestock, which are incidental in the effects of the mob which annually flits from Chinatown to the canneries."

By the beginning of the second week of August the alarm clocks rang for the last time, the canning was done and the workers embarked for Victoria. A 1920 report in the *British Colonist* described a group of Natives as they waited for the *Maquinna* to take them home:

THE GOOD HOPE CANNERY

> Within 2 blocks of luxurious Empress Hotel and the Provincial stately pile, has been for several days a little Indian village. On Belleville street, just beyond the Canadian Pacific Railway wharf, in a dozen or more primitive, squatty tents, the tent cloth pieced out by old quilts and blankets, about 75 Indian folks of all ages, have been living the simple life since last Sat. They are of Nootka, Hesquoit and Ahousat tribes, from the Ahousat dist, 150 miles up the West Coast, whither Maquinna takes them tomorrow, and have been fishing for the canneries at Rivers Inlet. Good fishing, too, the young men report, and good pay. Fifty-five cents apiece for sockeye runs into a lot of money in 6 weeks when the nets are pulling heavy. Weaving baskets at the tent door, the women conduct a bargain counter, which a few keen-scented buyers have found.

Native fishermen were at the mercy of the Federal Department of Fisheries and the canneries. Regulations passed in 1888 gave the Minister the power to determine the number of fishing boats and the type of gear used. Consequently, there were two kinds of licences, one issued by the cannery along with boat and gear, and one known as "independent." Natives were prohibited from having independent licences; only "British subjects" qualified. This policy was meant to encourage white settlement up the coast. Hoping to strike a balance, the government ordered the canneries to issue licences first to Native fishermen. Naturally, most cannery men no more liked losing control to independent white fishermen than they did being told whom they should favour when issuing licences. Bell-Irving stated the case for the cannery men to Fisheries officials in 1912:

> Imagine a fiord in Northern British Columbia running deep into the interior, between high mountains, the slopes of which are clothed with dense forest. No white man's habitation within a hundred miles, no human beings in the neighbourhood, except a few families of Indians, but at the proper season millions of Salmon. The [canneryman] arrives, and breaks the solitude. He clears a space ashore, erects his wharf and factory, and dwellings for his men, instals modern machinery and brings up materials, boats and nets. He then scours the country for labour, and soon what was once a solitude, is a busy hive of industry. His salmon pack is put up, representing a value of possibly $100,000 or more, and fish which before his arrival were

chiefly food for the bears and eagles, are converted into a valuable article of commerce. Later on the politician appears on the scene... He says to the canneryman—"It is true you have staked a lot of money in erecting and equipping your factory, but you must understand that after this the fish belong to the fishermen only, and you must buy fish from them at their price, though you will have the privilege as before of equipping them with boats and nets, and of advancing them cash and supplies when necessary." The canneryman thinks this is scarcely a square deal, but patiently continues. The large sums annually distributed in wages are of substantial benefit to the country in all branches of trade.

The Oowekeeno had no choice but to fish for the canneries under a rental agreement, making them salaried employees. But there was no guarantee that a cannery would issue them a licence. Tinshop George, the Oowekeeno chief and Bell-Irving's 1895 hunting guide, wrote a letter to BC Premier Richard McBride protesting a new manager at McTavish Cannery withholding fishing licences following a dispute:

> ... my people is in greatest trouble in our life. As not one of them get gear to catch fish this summer. ... The trouble is going by the new manager, the successor to G.S. McTavish. Mr. McTavish use to gave all the people of the Oweekaynoes their gears who wishes to fish here in previous season but we are in greatest trouble. We believe that it is not right to not suppo[rt] the Aborigines of Oweekayno River and we say since the first cannery planted in our richest place. And you can fairly well understand that most of my people are in middle of age and more older they only get their wages in fishing only. But after all that they could not get either job until another season come again. And it seems to us that we might get no more money to buy food or some of the people got crowd of family which I believe is going to be greatest trouble so I pray gave my humbly petition to you to try and help us out of trouble as we are your poorest Indian obedient friends.

In 1923 the federal government passed legislation allowing Natives to own licences. But the matter was hardly settled. In his *History of the Oowekeeno* David Stevenson writes, "The degree of independence was more in theory than in practice as many Indian fishermen

THE GOOD HOPE CANNERY

The first stop in the processing of salmon was Good Hope's "gutshed." Photo c.1940, courtesy of Mildred Dalton.

were forced to continue using company boats and company gear." He noted that "One Oowekeeno fisherman, John Thompson, explained how he had been misled into fishing as an 'independent.' He paid rent for the company gear and it was only in later years that he found out that the company gear or attached license made more money than he did as the gear was included rent free." In the face of discriminatory policies, unscrupulous practices, cultural assault and intense competition, the Oowekeeno persevered. "As early as 1913," wrote Stevenson, "there were three gasoline launches, as well as 12 fishing boats, plus canoes for a population of 108, listed by the Royal Commission on Indian Affairs, as belonging to the Oowekeeno."

Good Hope's population at the height of the 1911 season was 212 men, women and children, all living along a strip of rocky shoreline of just a few hundred yards. The families housed in a single row of identical shelters along the boardwalk. The Chinese housed—with their alarm clocks, presumably—in a large dormitory in behind the cannery itself, the Caucasians in dwellings on a slope to the southwest of the cannery and the Japanese at the opposite far end of the cove where a stream enters the sea. The energy, human and mechanical, must have been intense: hundreds of people at dozens of unceasing tasks; an endless stream of glistening silver sockeyes travelling up the ladder to the gut shed; cold stream water rushing into the cannery along Bell-Irving's flume; heat and smoke rising from the cordwood burning beneath the boiler; steam hissing from the retorts; hundreds of wheels spinning and belts whirring. And amid all the bustle were odours of fish, grease, oil, tin and lead solder; of sea water and human sweat; of burning wood; and somewhere, buried deep beneath the stink of human industry, the eternal base odours: damp, earthy forest perfumes and cold, salty sea aromas.

Other than news and mail delivered by ship, Good Hope stayed in touch with the world via a wireless telegraphy plant installed at the cannery around 1910, the first and the only one in Rivers Inlet for many years. As other canneries in the inlet shared Good Hope's wireless, privacy was maintained through a simple but elaborate cipher code. Every conceivable question or command about any aspect of ABC's operations was covered by this code. A single word starting with either an *R* or an *S* sufficed. Good Hope was *retenzione*, Phoenix *reticella* and Arrandale *retentatus*. If a cannery needed an engineer immediately it could cable *ricredersi*—"Send us a good gasoline engineer by first boat." If there was

Good Hope's power was supplied by a diesel engine manufactured by JB Petters & Sons of Yeovil, England. Photo c. 1940, courtesy of Mildred Dalton.

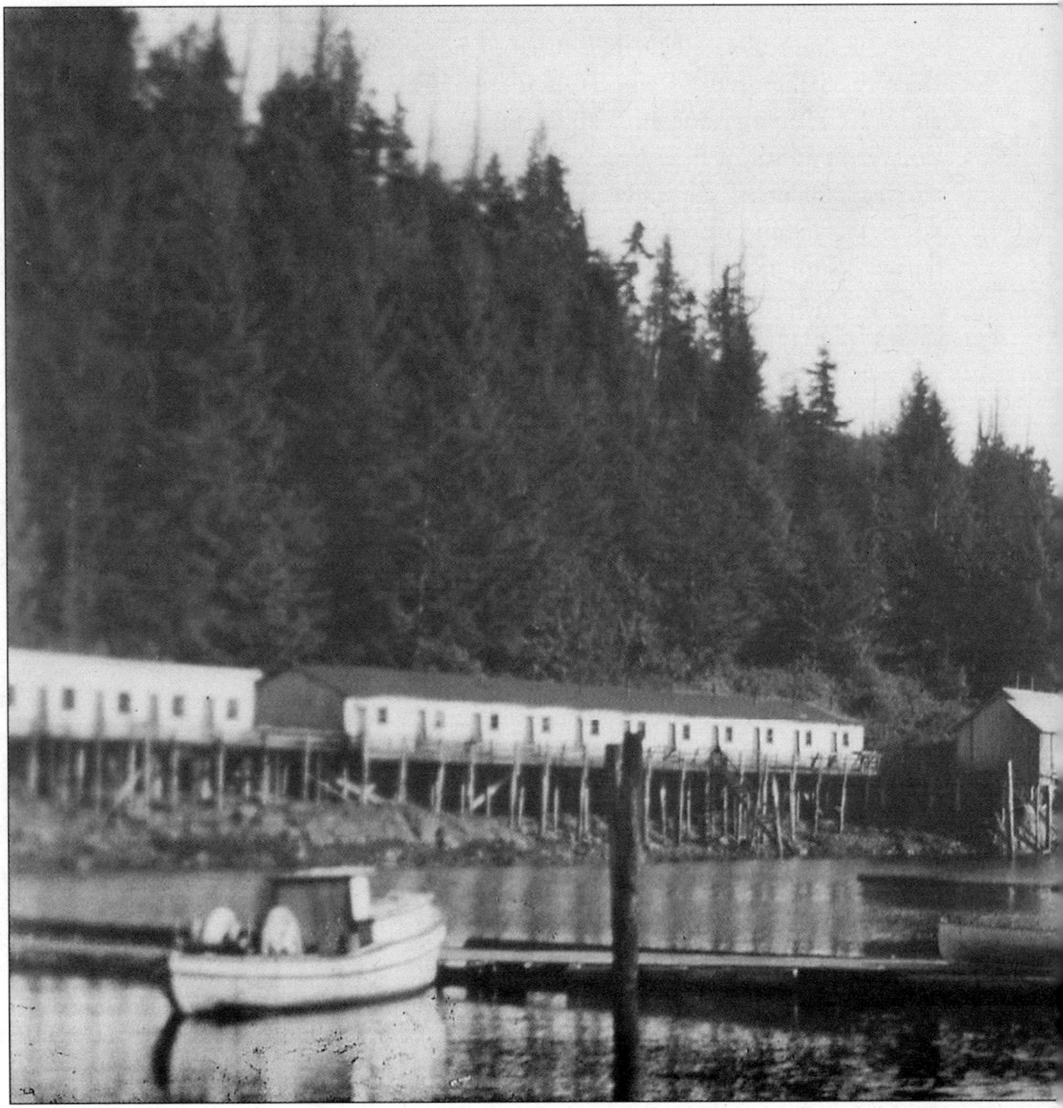

A quiet moment at Good Hope. The tranquility captured in this photo—and the absence of boats and people— suggests that the fishing season is either over or not yet underway. Housing on piles to left was for Native families, c. 1945. Photo courtesy of Maria Larsen.

THE GOOD HOPE CANNERY

At the outbreak of WWI Bell-Irving offered four ABC packer boats to the British Navy to assist in the defense of the BC coast. British sailors are seen here loading a torpedo on board the *Holly Leaf*, c.1914. Photo courtesy of Patricia Wilson.

labour turmoil, out went *ridamare*—"There is a strike of fishermen." Head office inquiries about pack costs were covered by *ributtato*—"At what actual cost can you pack?" The code for "How is the salmon run on Rivers Inlet?" was *scabbard*, and *sagebock* meant "Expect there will be a large pack on Rivers Inlet." *Sandront* conveyed, "The price of sockeye has been fixed at____." The code for the Bell-Irving head office was *rigesitto* and Bell-Irving's personal code was *rigging*. The code even covered each species of salmon, the fishing area and the four sizes of cans. Rivers Inlet sockeye in talls, for example, was *santessa* while coho in flats was *sarchiella*.

But there was no code for war. At the outbreak of hostilities in August 1914, the powerful, modern German warships *Leipzig* and *Nurnberg* were off the coast of California. The British had relinquished the Pacific Ocean and, in theory, Japan, a British ally, secured British interests through its Imperial Navy. But this was no comfort to British Columbians. The potential threat to Canada's West Coast was not theoretical; it was very real. British shipping on the west coast from Vancouver to the Panama Canal—opened in 1914—was paralyzed by the German naval threat. Bell-Irving had seen it coming. Since at least 1909 he and other members of the Vancouver Board of Trade had been pushing hard for public support in raising funds to make a contribution to the Imperial Navy. In 1910 Bell-Irving attended a meeting in Victoria at which Premier McBride passed a resolution favouring a Canadian Navy. Regardless, years of debate in Ottawa about the scope and mission of a Canadian Navy had yielded few tangible results. All that stood in the way of enemy warships were two antiquated sloops, the *Shearwater* and the *Algerine*, and one 1890s-era light cruiser, the HMCS *Rainbow*, presented by the British Admiralty to the infant Royal Canadian Navy (RCN) in 1910. Used for training, ceremonial visits and fisheries patrols, the 3,600-ton *Rainbow* could accommodate a crew of 273 and was equipped with two 6-inch guns, six 4.7-inch guns, and eight 6-pounder guns, but the vessel was too old and too slow for modern naval warfare. That did not, however, prevent the ship from sailing south from Esquimalt under Commander Walter Hose in early August 1914 with orders to find, engage and preferably destroy the *Leipzig* and *Nurnberg*. The *Rainbow* did not meet either of the German cruisers, although it missed encountering the *Leipzig* off the coast of San Francisco by only a day. To counter the German threat, Premier McBride purchased two Seattle-made submarines, originally intended for the Chilean navy, at $1.15 million for the pair, twice the annual budget of the RCN. Subs were stealthy and their torpedoes packed a wallop more than capable of sinking a warship. Victoria was defended by the large guns at Esquimalt. To defend Vancouver, the federal government installed two 4-inch calibre guns in Stanley Park and long-range field guns on Point Grey. But in the end, the coast of British Columbia saw no naval action.

After manoeuvres in the Alert Bay area, the torpedo-rigged packers stopped in at Good Hope on their way to Prince Rupert. Photo courtesy of Patricia Wilson.

The Bell-Irvings mobilized as well. In 1910 Henry Sr. had been instrumental in the establishment of Vancouver's 72nd Seaforth Highlanders. In August 1914, two of his sons, Roderick and Duncan Bell-Irving, went with the Highlanders to Valcartier, Quebec. Now it was his eldest son's turn to make a contribution. Henry Beattie Bell-Irving was the 27-year-old manager of H. Bell-Irving & Company and a member of the Royal Canadian Naval Volunteer Reserve.

THE GOOD HOPE CANNERY

At the end of August the British Bristol-class cruiser HMS *Newcastle* arrived in Esquimalt. Her commander, Captain Frederick A. Powlett, R.N., was senior to the *Rainbow*'s Walter Hose, and took over operational command at Esquimalt. Powlett prepared to lay mines to block off the northern passage between Malcolm Island and Suquash, Vancouver Island. To this end he commandeered the lighthouse tender *Newington*. Bell-Irving offered the Royal Navy four ABC fish packers, the *Ivy Leaf, Laurel Leaf, Holly Leaf* and *Viner*. His offer was accepted and the four packers, under the command of sub-lieutenant Henry Jr., were fitted out with dropping gear for torpedoes. A number of seamen for the new "torpedo boats" were drawn from the *Newcastle*. Bell-Irving noted in his diary that a man named Merrian captained the *Holly Leaf*, with Murray as engineer and J.H. Roberts as cook. On October 5 the *Ivy Leaf* and *Laurel Leaf* left for Alert Bay. The boats returned a few days later, but on the tenth once again departed for manoeuvres in the Alert Bay area along with the *Holly Leaf* and the *Viner*. Vice-Admiral Sir Charles Kingsmill, Canada's director of naval service, objected to a British commander making decisions over a sovereign Canada, and sent out a more senior naval officer who overruled Powlett's mine-laying activities.

With Canada ramping up for war, the fishing industry's material costs increased immediately. The price of flax gillnetting, for example, shot up 25 cents a pound. When the boats of the big international shippers Harrison Line and Blue Funnel were withdrawn from commercial service to aid in the war effort, shipment of the 1914 pack to Great Britain became problematic, and to European Allied ports impossible. The banks balked at extending credit. Canned salmon sat in Vancouver warehouses, adding storage, insurance and bank interest charges to the cost of production. As almost three-quarters of BC's canned salmon was exported, the canneries had a serious problem on their hands.

In December Henry Jr. received his commission and the four packer/torpedo boats were directed to go to Canoe Pass, near Prince Rupert, a port and railway terminus of strategic value. Not until enemy boats in the Atlantic and North Sea were accounted for and the German menace in the Pacific and South Atlantic was eliminated at the Battle of the Falkland Islands in December 1914, did the shipping routes reopen and the torpedo boats return home from the north coast. German submarine activity, however, continued throughout the war, limiting commercial shipping, doubling freight rates and increasing freight insurance costs. Henry Jr. stayed with the navy, serving in the waters surrounding Great Britain, and by war's end would earn the Distinguished Service Cross with bar.

As the war dragged on, demand for canned salmon increased. Not only was it protein-rich, but it was non-perishable even under the worst conditions and easily transported. The market price for canned salmon nearly tripled between 1913 and 1920. A 48-pound (22-kilogram) case of sockeye brought an opening price of $7.75 in 1913, $16 in 1917 and 1918, and over $20 in 1920. BC's canneries attempted to catch and can every available fish. All salmon species were now sought after. New canneries joined those already established, including—in Rivers Inlet—Provincial in 1917 and McTavish in 1918. In 1917 the fishing season at Good Hope and throughout the industry was exceptionally difficult. There were problems in obtaining supplies, and labour was scarce. Not only had three years of war decimated the ranks of experienced fishermen, shore workers and office employees but, attracted by offers of higher wages in other industries, Japanese and Chinese workers left the fishing industry. Despite these issues the 1918 pack in Rivers Inlet was higher than in the previous two years and the BC total salmon pack set a new record: 1,616,157 cases. Higher prices, however, were offset by higher costs and did not necessarily equate with higher profits for all canneries.

But that's not how Jack MacRae saw it. MacRae, the fictional hero of BC author Bertrand Sinclair's 1920 novel *Poor Man's Rock*, is a World War I veteran who challenges the canneries by leasing packer boats and paying higher prices than the canneries for fish, putting more money in the pockets of fishermen by taking a smaller cut for himself:

> The fishermen had made a living, such as it was. The cannery men had dwelt in peace and amity with one another. They had their own loosely knit organization, held together by the ties of financial interest. They sat behind mahogany desks and set the price of salmon to the fishermen and very largely the price of canned fish to the consumer, and their most arduous labor had been to tot up the comfortable balance after each season's operations. All this

pleasantness was to be done away with, they mourned. Every Tom, Dick, and Harry was to be turned loose on the salmon with deadly gear and greedy intent to exterminate a valuable species of fish and wipe out a thriving industry. The salmon would all be killed off, so did the packers cry. What few small voices arose, suggesting that the deadly purse seine had never been considered deadly when only canneries had been permitted to use such gear and that *they* had not worried about the extermination of the salmon so long as they did the exterminating themselves and found it highly profitable—these few voices, alas, arose only in minor strains and were for the most part drowned by the anvil chorus of the cannery men.

A wealthy canneryman character surnamed Robbin-Steele calls MacRae into his office:

MacRae faced the man over a broad table in an office more like the library of a well-appointed home than a place of calculated profit-mongering. Robbin-Steele, Senior, was tall, thin, sixty years of age, sandy-haired, with a high, arched nose. His eyes, MacRae thought, were disagreeably like the eyes of a dead fish, lusterless and sunken; a cold man with a suave manner seeking his own advantage. Robbin-Steele was a Scotchman of tolerably good family who had come to British Columbia with an inherited fortune and made that fortune grow to vast proportions in the salmon trade. He had two pretty and clever daughters, and three of his sons had been notable fighters overseas. MacRae knew them all, liked them well enough. But he had never come much in contact with the head of the family. What he had seen of Robbin-Steele, Senior, gave him the impression of cold, calculating power.

Robbin-Steele tries to get MacRae to guarantee him a supply of fish at a set price:

"You are laboring under the common error about cannery profits," Robbins-Steele declared pointedly. "Considering the capital invested, the total of the pack, the risk and uncertainty of the business, our returns are not excessive."

"That all depends on what you regard as excessive... Canning salmon is a highly profitable business, but it would not be the gold mine it has been if canneries hadn't been fostered at the expense of the men who actually catch the fish, if the government hadn't bestowed upon cannery men the gift of a strangle hold on the salmon grounds, and license privileges that gave them absolute control. I haven't any quarrel with cannery men making money. You only amuse me when you speak of doubtful returns. I wish I could have your cinch for a season or two."

After more verbal sparring, MacRae declines Robbin-Steele's gambit:

"You are a very shrewd young man, I should say." Robbin-Steele paid him a reluctant compliment and let a gleam of appreciation flicker in his dead-fish eyes.

A working west coast trollerman and a veteran pulp fiction author, Sinclair lived the life he wrote about. *Poor Man's Rock* is an entertaining story that also happens to provide non-fictional, first-hand insight into the issues and tensions of the fishing industry of the time. But for the student of H.O. Bell-Irving the novel yields a secret pleasure, because Sinclair was obviously thinking of him when he wrote it. The similarity is more than just the hyphenated surnames. Like Robbin-Steele, Bell-Irving was a fit, sandy-haired Scotsman of sixty years who had come to Canada to make a fortune and had sired clever daughters and had sons who were "notable fighters overseas." It is highly unlikely that Bell-Irving and Sinclair ever met, although Sinclair did live in Vancouver on and off for a number of years at the same time as Bell-Irving. But he picked the right model for his novel's antagonist. Profitwise, the ABC company did very well through the war years. In 1919 Bell-Irving noted in his journal how much money the company made from 1912 through 1918:

1912	£67,892
1913	£12,848
1914	£14,402
1915	£52,245
1916	£35,736
1917	£68,794
1918	£177,246

THE GOOD HOPE CANNERY

In July 1920, the same year that *Poor Man's Rock* was published, Bell-Irving was at Good Hope. Inspecting the can loft, he noticed rust on some cans. Rain was the culprit and he noted that the roof over the can lofts and part of the net loft needed re-shingling. He also noted, much to his displeasure, that a new scheme was in play, one worthy of Jack MacRae. What roused Bell-Irving's ire was a "racket" started, so he claimed, by Kildalla Cannery's manager Dan Groves. The Indian and Japanese fishermen, after catching 600, were, a la MacRae, to "get independent prices!" Of Kildalla's 82 boat licences, 80 boats were fishing, comprising 41 white fishermen, 30 Indians and 11 Japanese. Of those only 13 furnished their own nets and only one his own boat. And there was "only one Indian owned boat in the whole lot!" Furthermore, a 21-year-old Indian was the high boat with 2,500 fish. Maybe Groves, with only 13 Japanese fishermen and relatively few Indians compared to ABC, could afford to pay independent prices, but damned if Bell-Irving was going to.

Dr. R. Geddes Large, son of Rivers Inlet pioneering doctor R.W. Large, describes in *Drums and Scalpel* the Rivers Inlet's Fishery Department regulators and their vessels circa 1910.

> The fishery patrol vessel was a trim little boat with a Buffalo engine. Called the *Merlin* and operated by Gunnar Saugstad from Bella Coola… the *Falcon* [was] the Fishery Department's patrol vessel for the north coast, on which the District supervisor traveled, overseeing the work of his inspectors on the various fishing grounds. She was a beautiful boat about 60 ft. long and kept in spotless condition—painted white with varnished decks. Mr. Williams, the District Supervisor, was a fine figure of a man with a waxed mustache and always dressed in navy dress uniform with white flannel trousers and white shoes. He made a very imposing figure which undoubtedly lent dignity to the law which he enforced in his capacity as a Stipendiary Magistrate.

Having reserved skiffs for the season, the fishermen arrived at Good Hope a week before the opening. Their skiffs were already in the water waiting for them, freshly limed white inside and painted ABC red on the hull. Andy Anderson, son of Good Hope manager Oly Anderson, tells a story about the last fisherman to ever fish from a skiff at Good Hope:

He asked my Dad for a can of red paint to spruce up the old boat. When he had it done he told Oly that he was going out to do some fishing. "But the paint's still wet," said Oly. "You better wait a day for it to dry." "No, I can't wait," and he sat down in the skiff and started rowing, giving the rear end of his pants a nice red glow for the rest of the season.

After bailing out any rainwater, the fishermen set up their cotton shelters forward of the main bulkheads and stowed away stoves, bedding and a week's worth of food.

One early fisherman described life aboard a skiff this way: "The sleeping quarters were simply a couple of blankets on the hard bare board on which we lay, crouched under the foc'sle head which was not long enough to cover our entire body, thus leaving our legs exposed, over which we spread our yellow oilskin raincoat. Our cooking was done on a five-gallon coal oil can which was cut down to form a crude stove in which we used wood for fuel. Our lavatory? Well, we sat over the boat side. In rough weather we tied ourselves to the mast to avoid being washed overboard."

On opening day the skiffs hitched rides to the fishing grounds on Good Hope's steam-powered, and later gas-powered, towboat. Lester Peterson in "Fishing Rivers Inlet By Sail and Oar" describes the hitching up and unhitching process:

In Rivers Inlet, skiffs were not linked one to another, as was done in other places along the coast. Here, each tow-boat was equipped with a heavy line. At skiff-length intervals [about 25 feet/7.5 metres], smaller lines, about four feet in length, and with hardwood eyes set in the loose ends, branched out alternately from either side of the main line. Each fisherman received one of these branch lines. He fed his painter rope through the eyes and tied the end in a bow knot amidships. With the latest arrival nearest its stern, the tow-boat set out toward its destination, pulling as many as forty skiffs. A fisherman had no contact with the tow-boat, sometimes far ahead of him. When he wanted to drop-off, he was obliged to follow a very exact and strenuous procedure. First, seated, well-braced, he pulled undone his painter and held it by the end. Then, with the other hand, he grasped the tow-line, which lay, rigid from the pull on it, alongside his skiff. Holding to this, and traveling full speed, he gradually let out his painter, through the eye of his branch line, allowing the bow of his skiff to veer away from

THE GOOD HOPE CANNERY

Skiffs being towed out of Good Hope, Ethel Island, can be seen in the distance. 1940s. Photo courtesy of Marea Larsen.

the tow-line. Selecting his own "moment of truth," he released, first painter, then tow-line, at the same time beseeching his private gods to keep him clear of the double column of vessels charging at him from astern.

Fisherman Clayton Mack, writing of his time fishing in the 1940s, gives us his take on being towed to the fishing grounds: "Wind blow too hard in, they have to tow us out to go fishing. Maybe twenty boats on one towline. Look nice, you know. All the masts in a row behind that tugboat. Tugboat start out around twelve o'clock after lunch. Tow us out to where we want to go. Lot of fun. Just let go of the rope, pull the sail up and go where you want to go."

It was up to the manager to make sure that the skiffs were stationed where the sockeye were likely to be. In the early weeks of the season the fishermen gathered at the mouth of the inlet. They worked in Schooner Passage (Darby Channel) between Walbran Island and the mainland before moving toward the head of the inlet where Shotbolt Bay was the most desirable fishing ground. As strong as the inlet's tides were, rising 16 to 20 feet (5 to 6 metres), they were not the main determinant of fishing success; that distinction went to the weather. When it rained, the salmon stayed deep and slipped under the gillnets.

Bertrand Sinclair describes making a set: "A gill net goes out over a boat's stern. When it is strung it stands in the sea like a tennis net across a court, a web nine hundred feet long, twenty feet deep, its upper edge held afloat by corks, its lower sunk by lead weights spaced closed together. The outer end is buoyed to a float which carries a flag and a lantern; the inner is fast to the bitts of the launch. Thus set, and set in the evening, since salmon can only be taken by the gills in the dark, fisherman, launch, and net drift with the changing tides till dawn. Then he hauls. He may have ten salmon, or a hundred, or treble that. He may have none, and the web be torn by sharks and fouled heavy with worthless dogfish."

 Having set their nets and, as darkness neared, replaced the company flags with lanterns, the fishermen in Good Hope's early days settled in for the night and a few hours of sleep. As the sun came up they got into hip boots and oilskin aprons. If the weather was bad they threw on rubber coats and sou'wester hats. In the unwanted company of sockeye-stealing seals, 80-foot (24-metre) sperm whales and killer whale pods, they picked up their nets over a stern roller, splitting the lines as they came in, cork line forward and lead line behind their boots. The sockeye, caught by their gills, were freed from the mesh. After bailing out, the fishermen rowed clear of others and set their nets again. When the collector boat arrived later in the morning, the fishermen pitched their sockeye aboard and the collector's skipper recorded the catches in their tally books. The fishermen ate, rested and looked after their nets until it was time for the next set. They repeated the process day after day until Friday arrived and the collector boat towed them back to the cannery. Here they hauled their nets into large wooden tanks holding an algae-dissolving solution of copper sulphate known as "bluestone." A net man armed with a needle mended any holes torn in the mesh. The fishermen then descended on the cannery store for supplies, grabbed showers and shot the breeze with each other. Most slept on board their boats. When at last the Rivers Inlet season ended, some of the fishermen fished into the fall on seine boats, others resumed logging or trapping, and some went back to their small farms or "stump ranches," taking on odd jobs to help make ends meet. His cannery days over, Good Hope veteran millwright engineer Robert Jamieson added to his income by making and selling wooden bric-a-brac, pipe holders and kitchen towel racks.

~

At the request of cannery operators, gasoline-powered boats were not allowed to carry nets in Rivers Inlet until 1924. Ostensibly a fish stock conservation measure, permitting the use

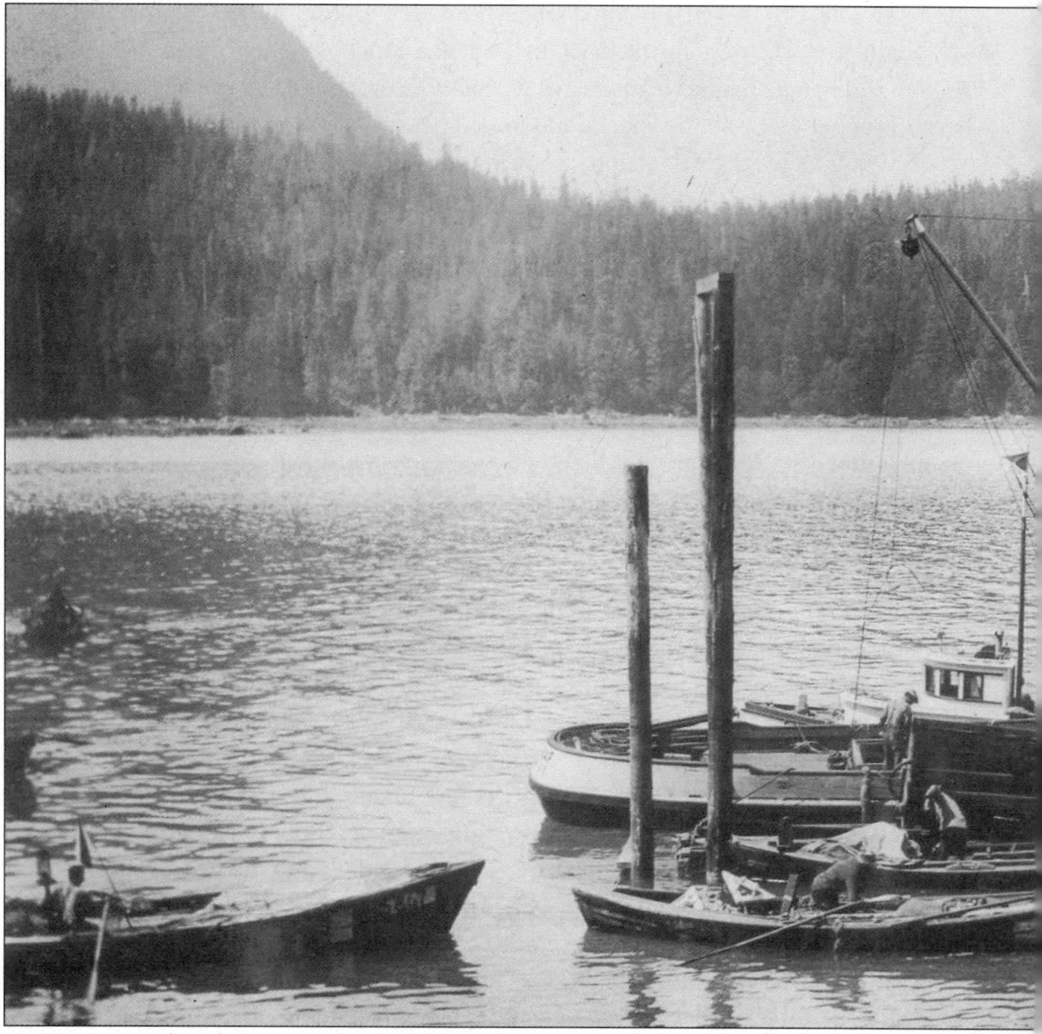

Anglo-British Columbia Packing Company's *Pine Leaf* unloading fish at Good Hope, 1930s. Note the ABC flag and fish ladder leading to the gutshed. Photo courtesy of Marlene Yurichuk.

THE GOOD HOPE CANNERY

of gas boats would have altered the 1903–4 boat-rating system, lowering the number of boats each cannery could put out fishing and consequently the number of fish they could catch. That wasn't about to happen, and so for the first twenty-nine years of its existence Good Hope's fishermen had no choice but to fish from cannery-owned skiffs. Rivers Inlet Cannery had the lion's share at 114, about a quarter of all skiffs allowed in the inlet, followed by Wadham's with 88, Good Hope with 76, Brunswick with 69, Wannock with 57 and Vancouver with 54. The boat rating for Good Hope was increased to 100 in 1908 and reduced to 95 in 1910. When gas boats were finally permitted, only Europeans and Indians were allowed to use them, a measure to placate the two racial groups at the expense of the ever-expendable Japanese. Still, many fishermen continued to use skiffs into the 1950s.

By 1936 gas boats were in use in Rivers Inlet and cannery operators set a price of 45 cents a sockeye, except in the inlet itself where, because of its smaller sockeye, the price was set at 40 cents. After fishing for a week, the Rivers Inlet fishermen went on strike, demanding 50 cents a sockeye. Neither side would budge. Negotiations failed and the fishermen would not consent to arbitration. The season was a loss for all concerned. They were back in business in 1937. Isabel Edwards, in *Ruffles On My Longjohns*, describes the action that year at the head of the Inlet:

> At six o'clock [Sunday night] the signal boomed and pandemonium broke loose. Boats roared around seeking a place, sideswiping each other, while men shouted and cursed above the din of the engine noises mixed with the roar of wooden corks banging like machine-gun fire as they ran out across the gunwale in a frenzy of speed. It was a madhouse and in the hysteria, men corked each other by throwing their nets too close, or even across someone else's net. One young chap in an open power boat ran amuck with his throttle wide open, dashing aimlessly in and out among the other boats, scraping their hulls and running over nets… after dark, lights were everywhere in utter confusion… there was no sleep for anyone that night but by morning the hysteria had died away… bleary eyed men were pulling in their nets and counting the catch.

CATCH WHAT YOU CAN, CAN WHAT YOU CATCH

World War II brought changes to the marketing of BC's canned fish as the industry and its important protein-rich product gradually came under federal government control. Great Britain received approximately 70 percent of BC's 1941 canned salmon, and from 1942 to 1946 the entire fisheries production of BC was under government allocation. Of the 1941 to 1946 packs, 7.6 million cases (80 percent) were used for the armed forces or shipped to the order of Great Britain for overseas needs. With a ready market, guaranteed sales, set prices and controlled costs, the war years were profitable for salmon canners.

Fishermen with their own gas-powered gillnetters journeyed from all points to Rivers Inlet and Good Hope. Here Albert Ericson, his daughter and John Grohn enjoy a meal on the "Rolling On," Johnstone Strait, c. 1934. Photo courtesy of Marty Grohn.

Hugh McKervill, in *The Salmon People*, describes fishing in Rivers Inlet in the 1940s:

During July the main run of sockeye arrives and excitement reaches its peak. Fish-fever drove the fleet in its wild search for the bonanza catch which every man hoped for but few managed to find. Up at the head boats pressed close to the boundary imposed by the Fisheries Department. On the falling tide they lined up and played out their nets between the white triangles that marked the boundary on either shore. Tempers wore thin in jockeying for position. Nets tangled together, propellers snagged, swearing in a dozen accents could be heard above the sputtering of motors and if the fishery guardian happened to be Ray LaMarsh a boat that drifted across the line was liable to be rammed. For Ray had his own theories about

THE GOOD HOPE CANNERY

how to keep fishermen out of the prohibited area where fish were supposed to be free from the menace of nets.

Charlie Lord was Justice of the Peace in matters pertaining to Fisheries between the years 1932 and 1947. The old Indians liked and respected Charlie because he could speak Chinook and always explained the formidable legal language to them. He was a working man at heart and hated to impose fines on fishermen whom he knew needed every cent they could get. But he was also a Britisher with a strong sense of the dignity of the law. Many fishermen were

Good Hope's fuel dock was located across the bay from the cannery, as a safety precaution. Photo courtesy of Marea Larsen, c. 1940.

arraigned before him down by Dawson's Landing on a Saturday morning because their nets had drifted into restricted water.

From the boundary, past Round Mountain, into Moses Inlet, Brunswick Bay and by the old hospital now empty, gaunt and green with moss, the gillnetters drifted. They fished through Schooner Pass, so narrow that nets had to be woven across in a series of S's. Around Welch Island and in every nook and cranny of Darby Channel, out past ominous Swan Rock and onto Fitzhugh Sound the boats were strewn. They drifted off Long Point and slid down the

Accommodation was available to fishermen in the cabins lining the shore (right.) A boardwalk hugging the shore connected these buildings to the cannery. Photo courtesy of Marea Larsen, c. 1948.

Inlet with the falling tide. Sometimes a sleeping fisherman would awaken to discover his net had wrapped around the Haystack, a tiny island named for its shape. Further down an unwary gillnetter could get sucked into skookum-chuck at Draney Inlet when the tide flooded. Out past Goose Bay to Major Brown Rock when the Pacific swells pounded and washed at the bald stone outcroppings with eternal determination the boats reared and plunged in the waves, or bobbed lazily during a rare calm.

Once gasoline engines were introduced there were boats of all descriptions and sizes, ranging from "high-liners" rigged with the latest equipment and bristling with paint, to tired old tubs that barely stayed afloat. There were Japanese, Indians, Swedes, Czechoslovakians, Scots, Finns and Greeks. There were teenage boys working furiously, nervously. There were old salts whose rhythmic movements were not squandered on unnecessary labour. And for a few seasons two Vancouver girls were afloat on the Inlet. Their net seldom touched the water but each morning they would have a box full of fish to deliver, and it became general knowledge that the price of a nocturnal visit to their craft ran as high as twenty-five prime sockeyes.

If the Vancouver girls were welcome, other inhabitants of the inlet were not. In a 1941 edition of *The Fisherman* newspaper, the writer expresses the prevailing wisdom on fish conservation measures: "Sea lions were very much in evidence, tearing nets and devouring fish. It has been suggested that the Fishery Dept. co-operate with the air-force and bomb every rookery on the coast. It would not only be good practice, but would also fulfill a much-needed want." For the fishermen, there were just never enough fish. "Fishing in Rivers Inlet has not been very startling this year," lamented *The Fisherman* in 1942. "The average has been about 600 fish per boat up to July 25… somewhat lower than the normal average… high boat was around 1,200 with low boat around 375. Sockeye were large and weighed an average of 6 lbs per fish." The situation was even worse in 1943. On July 20 *The Fisherman* reported that "Sockeye have been scarce so far. Perhaps weather has something to do with it. We have been having real November weather, so much so, that some of us are starting to worry about buying Christmas presents."

There were plenty of accounts of fishing in Rivers Inlet from white fishermen, but where were the Native accounts? For the longest time I couldn't find any, but one day at the Semiahmoo

CATCH WHAT YOU CAN, CAN WHAT YOU CATCH

Library in White Rock I came across *Living On The Edge: Nuu-Chah-Nulth History from an Ahousaht Chief's Perspective* by Chief Earl Maquinna George. Chief George was the first hereditary chief of the Ahousaht First Nation of Clayoquot Sound. Born in 1926 in Maaqtusiis on Flores Island, he spent his early years at the Ahousaht Indian Residential School as his mother had died when he was a child. In 1938 he accompanied his father, McPherson George, to Good Hope Cannery. His experience comes one year before Good Hope absorbed McTavish and two years before Good Hope stopped canning.

No room to spare. Native passengers and cargo piled high on a ship's foredeck. Rivers Inlet, c. 1926. Photo courtesy of Barb Little.

THE GOOD HOPE CANNERY

During the early part of my life in the 1930s and early 1940s, I recall in the early summers the people were home in Ahousaht, and June saw the beginning of the calm and beautiful early summer days. The summer holidays usually started on June 18 or 19, when the children used to go home from the residential schools to their parents for the summer months. Around that week the adults usually started to move towards the fish canneries located up and down the coast. The Ahousaht usually went up to the Rivers Inlet area called Oweekeno, where there were many fish canneries. My father had a little boat called *The Native Lass*. It was made of cedar, was about 30 feet long, a combination gillnet-troller, powered by a single-cylinder, a "one-lunger." My dad finished the engine overhaul, and the middle of June came pretty fast, when we were all released from the residential school for the summer. So, we went, stopping in places like Hotspring Cove in Clayoquot Sound, Friendly Cove, Nootka Sound—various other villages. We anchored out in places like Nuchatlitz, Queen's Cove, and in the mornings we left early… we had a boatload of people who were going up with us to work in the cannery. There were at least 12 people…

The journey to the Rivers Inlet fish canneries could be very dangerous for the small boats traveling up the coast. The travelers put a canvas cover over from the cabin to the stern of the boat for the trips. That is where the ladies and children had protection from the cold and rain. The environment was always an uncertainty, but our people were familiar with the anchorages up and down the coast that they could use for shelter. Although beautiful in the summer, the sea can be treacherous with its winds and tidal streams. When the tides are strong they rise up to 12 or 13 feet, and the current runs offshore along the outer side of places like Estevan Point at the tip of the Hesquiat Peninsula. Those conditions can be perilous for a small boat taking water into the hold with the weight of the passengers, as many as 20 in the family, and often several families.

We went into Rivers Inlet after traveling around Cape Scott on the northern tip of Vancouver Island, and then across the way to Rivers Inlet to the place called Good Hope Cannery. My dad said, "You should apply for work. You might make a few dollars in the fish cannery," and I agreed. So we went to see the Chinese boss, who was taking in the applications of those who wanted to work in the fish cannery. The boss said, "We start you off at 15 cents an hour, which is not very much, but that's the pay the Chinese labourer gets, and that's the pay you also will be getting…" There were other young people who were hired on in the cannery, but many quit as soon as they saw what kind of work they had to do, and also how much pay they would be getting. I thought about quitting on the first day too. It seemed like a very long day, 17 hours on the first day of my job in the fish cannery. It was very demanding work. I had to sit at the end of a chute, just big enough for the cans to roll down from the upper part of the second story of the building, where Chinese workers were rolling the cans down. I waited until the basket that held the cans was full and then quickly pulled the filled basket away and put an empty one in its place. The baskets full of cans were brought to the ladies who were packing the cans. And when each can was filled, the lady put it on a conveyor belt and all the cans moved along the line of the conveyor belt, held in a tray for them to go through the steamer.

The ladies were also under contract to the Chinese boss. The way they were paid was piecemeal for each small wooden tray they filled. There was a Chinese worker that had a hole puncher, and the lady had a coupon hanging over where she was filling the cans. The coupon was the way they tallied the number of fish the lady had processed each day. Not all the ladies got the same amount of pay. Some were quite fast, and others were quite slow in filling the cans. I remember the look of all the arms and hands moving when the fish were being put into the cans. Once they were in the cans, the fish went onto another conveyor belt and into a steam box that had a tray, and when the tray turned over, the fish were put through the steam box and moved along out to the other end, where they came out after being steamed.

I worked until the cannery closed… when the fishing season ended. At that time I went to the Chinese boss, and he wrote out my payroll statement. I handed my pay—13 dollars in cash—over to my aunt. She demanded that she would hold my money until she was ready to give it to me, when I needed something, to use it wisely. That was my aunt's purpose for taking my money. She tied it in a sock and hid it somewhere. I didn't know where she had it hidden, even though we were traveling in a small boat. Then we traveled back to our homeland, and eventually I went back to the residential school. I told my friends in the school that I had been working all summer and that I was broke already! My aunt let me spend my money buying a shirt and socks and underwear, which left me completely broke by the time I returned to school.

There was something familiar about the Chief's story, an echo of something else I'd heard or read. I turned it over in my mind, trying to remember where I'd come across it before and wondering too if Levi Lauritsen or Harold Wood were aware of the boy at the bottom of the can chute shuttling baskets hour after hour.

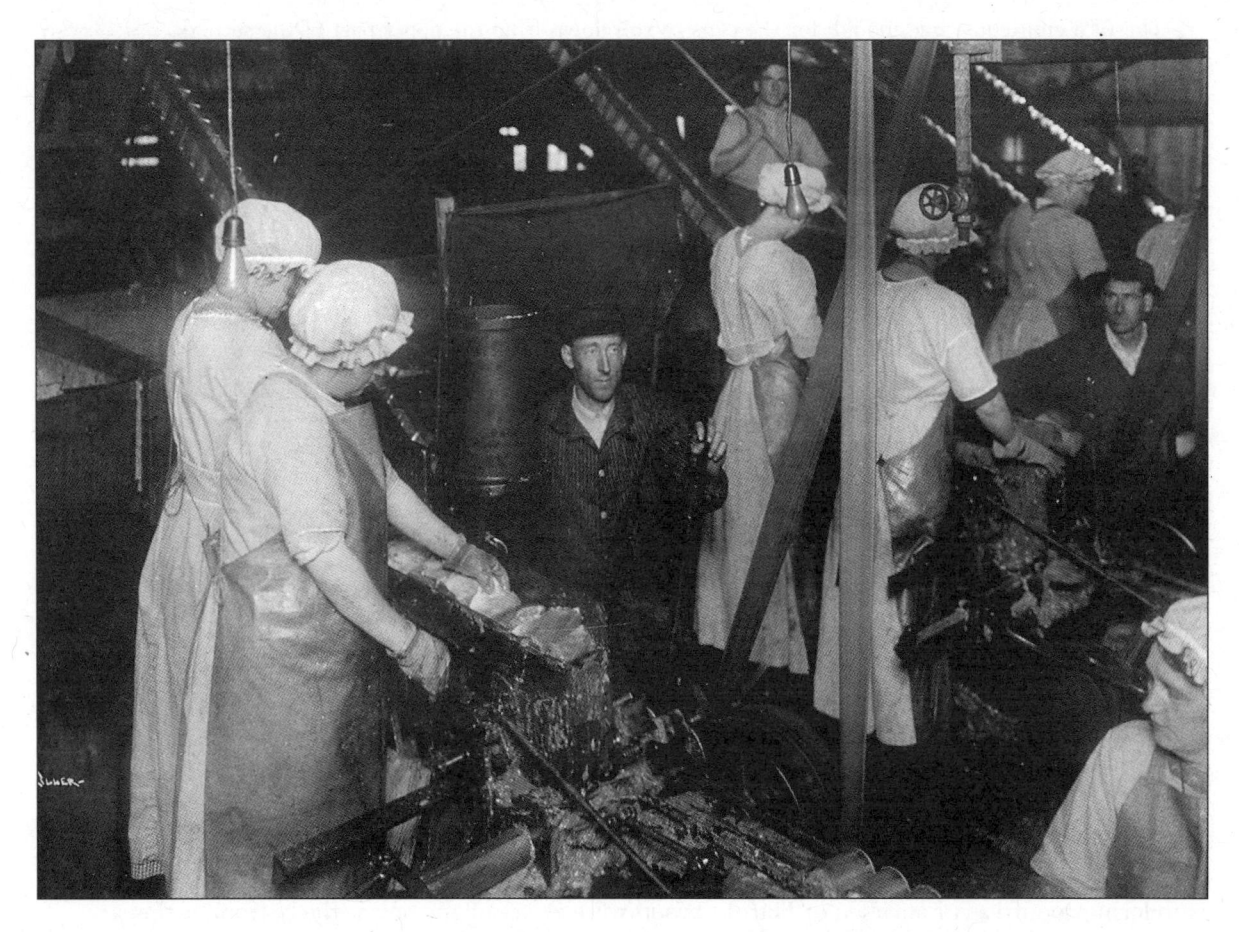

This photo taken at ABC's British American Cannery in Canoe Pass, Fraser River, shows the sludge and slime that was part of the everyday working conditions for a cannery worker, c. 1905. Photo courtesy of Ian Bell-Irving.

Washing salmon at the British American Cannery, Canoe Pass, c. 1905. Photo courtesy of Ian Bell-Irving.

The canning line at Good Hope ran from 1895 to 1940. Photo courtesy of Mildred Dalton, c. 1940

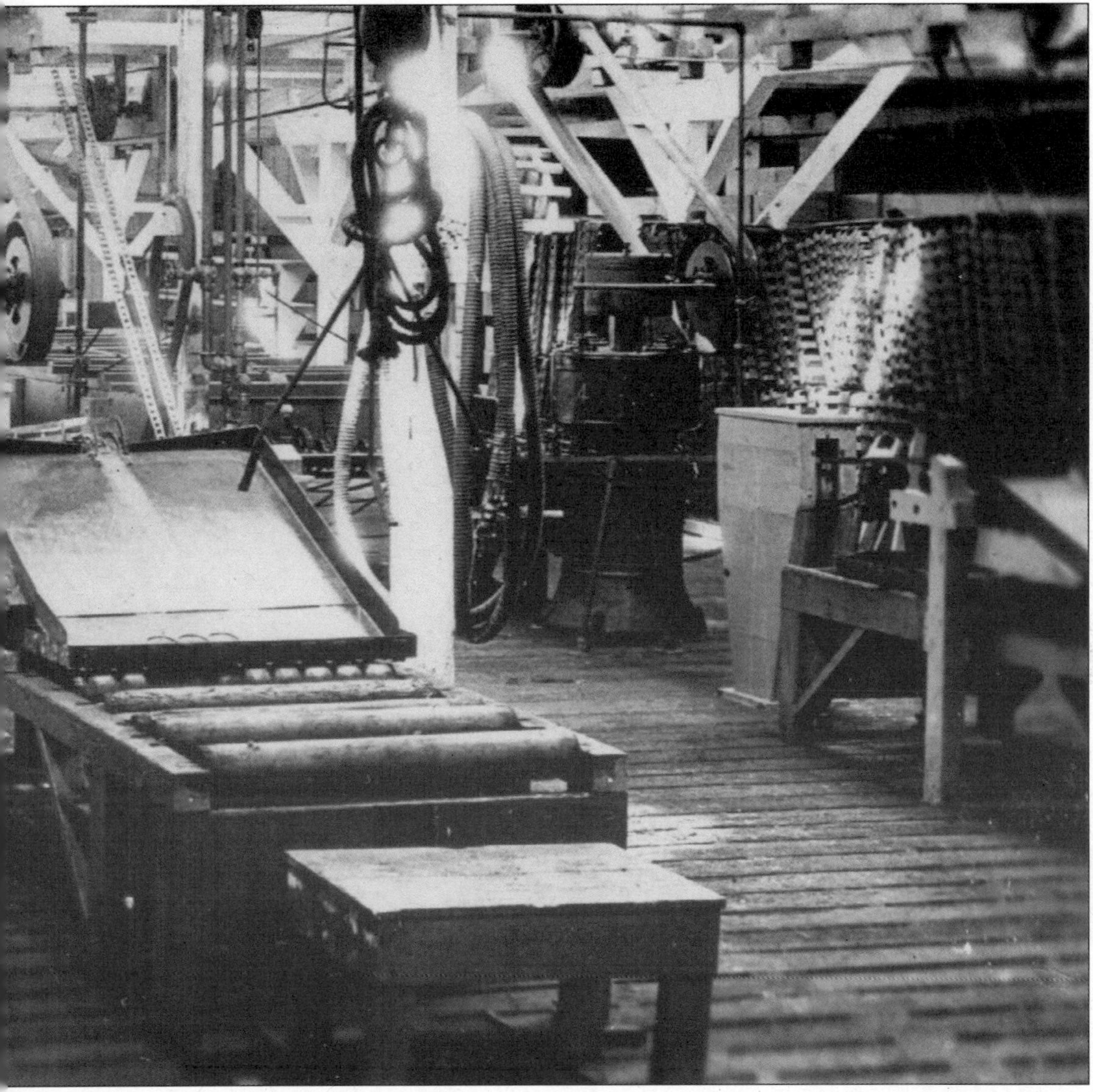

Salmon in cans waiting for lids. Like most jobs in the cannery it was messy work. British Columbia Cannery, c. 1905. Photo courtesy of Ian Bell-Irving.

Whole salmon carcasses being cut into steaks at the British American Cannery, Canoe Pass, c. 1905 Photo courtesy of Ian Bell-Irving.

Four-year-old Marea Abelson moved with her mother Irene and father Olav, a fisherman and winter watchman, to Good Hope in 1939. The Abelsons were permanent fixtures for the next eleven years. Photo courtesy of Marea Larsen.

Chapter 5

Marea's Story

Thor Ovitslund was a strong, good-looking, young guy. He was up at Sandell Lake with two other people. They were on shore and he was out on the lake in the canoe. He said to them, 'I'm taking off my boots, it's safer that way.' It was the last thing he said. He went through and the cold water got him. He could swim, but the water was just too cold. The two on shore couldn't swim.

—Mildred Lauritsen

Irene Brodie was born December 17, 1900, in Markdale, Ontario. She came to BC in the early 1920s to keep house for her Uncle Syd, a retired Ladner farmer who owned an acre lot in south Vancouver. After marrying Olav Andreas Abelson, she worked at the McTavish Cannery in Rivers Inlet in 1934. Olav was born in Kvale, Norway, on February 1, 1903, the eldest by eleven years of two sons. The Abelsons were farmers, carpenters and boat builders. Olav's father built fishing boats that were able to cut through the *saltstrummer*, the ocean's rapid currents, by virtue of their sharp-keeled design. He came to Canada in 1923 with a group of other young Norwegians from the same area, looking for a better life in the new world.

Olav met Irene six years later in Vancouver at the Silver Slipper Dance Hall, and one of their first dates was a hike up Grouse Mountain. They married on February 1, 1934. Olav fished for the McTavish Cannery and was the winter caretaker at Good Hope. Irene was a keen gardener and people visiting Good Hope would make a point of seeing her work, gardens being rare along the coast. The Abelsons' only child, Marea, was born in 1935. Marea

Abelson cottage, Good Hope, 1940s. Photo courtesy of Marea Larsen.

and Irene joined Olav at McTavish in the summer of 1938 and lived there until it closed down permanently in 1939. From then until 1950, they lived at Good Hope year-round.

As chance would have it, Marea was my brother Donald's ex-mother-in-law. One afternoon in June 2008, Donald and his 14-year-old son, Ryan, dropped by for a visit. I launched into my favourite topic and as I went on about the Good Hope Cannery and the records found in its walls, Ryan turned to me and quietly said, "My grandmother lived at Good Hope when she was a girl." I couldn't believe my luck! The odds were against me knowing anyone who had lived at Good Hope, let alone someone who had been there during the same era as the bookkeeping records. I hadn't spoken with Marea in years, but when Donald and Ryan's mother had been together I had known her and her husband, Terry, very well. I called her and we arranged to meet the following week at her home in Maple Ridge.

Her Story

In the fishing season we lived in a house at Good Hope just up the hill from the beach at the end of the bay. At the peak of the summer there were times when there were at least 200 people at the cannery, 150 fishermen, the crew that worked on shore, the net bosses, the storekeeper and helper, the bookkeeper, the machinist and, of course, the cook, and the manager. There were others as well, such as my father who would start work in the springtime getting

MAREA'S STORY

all the repair work done on the pilings and the boardwalks, [so] that everything was in shape for the fishing season. The sockeye season started at the end of June and went until the beginning of August, and it was a frenzied time. On a weekend when the fishery was closed there was a real hubbub at Good Hope, because the fishermen wanted to bluestone their nets, get a shower, freshen themselves up, do some laundry and buy supplies. The Indians had a little kick-a-poo joy juice going sometimes. They would get a gambling game going that they played with sticks.

Skiffs were still in use in the '30s and '40s when they fished from six o'clock Sunday until six o'clock Friday night. Depending on how good the fishing was, there would be a packer making a trip to the cannery at Steveston probably twice a week. [Good Hope stopped canning in 1940.] If it was a busy, busy time there might be three, but usually maybe Tuesdays and Fridays. Indian women were very happy to be out fishing with their husbands on a skiff. The Indians were very, very family oriented.

My mother made the flags [an 1899 ABC innovation] for the boats for a number

Irene Abelson's garden was a must-see for visitors to Good Hope. Olav, Marea and Irene pose amid flowers on the walkway to their cottage. Photo courtesy of Marea Larsen.

THE GOOD HOPE CANNERY

of years. Each gillnetter was issued three flags: one large one for the boat, and two to put on each end of the net. Each cannery had its own colours; ABC was red and white diagonals. She made the flags on a treadle sewing machine, but they were always late getting the material up; it never came on time. So here she was panicking as it got near the opening of the season because all the fabric had to be cut out in the proper shape and sewn by hand. As she made them I would run them over to the cannery office so that the fishermen could pick them up. One day an airmail letter came telling Mom that her father had died of a heart attack and, you know, my mother sat sewing those flags day after day because she had to get them done, and she cried. I can see her with the tears running as she sat sewing. Oh, it upset me so to see my

Marea and her feline friends. The cats and dogs were good company. Good Hope, early '40s. Photo courtesy of Marea Larsen.

mother crying because she never cried. For years after whenever I saw an airmail stamp on a letter it brought back memories of that.

I was a great daydreamer, always talking to myself and to my cats and dogs, and as far as I was concerned they communicated with me. I would row out to the island with my cocker spaniel, Lassie. I also had a terrier named Jiggs who liked to climb up the ladder onto the roof of our woodshed. Jiggs was given to me by a fisherman; he was too much of a rascal on board his boat. I had a rash on my back one time, I was six or seven. We couldn't figure out how I got it. Dr. Darby came and had a look at it. I'd gotten it from being on the roof of the woodshed with Jiggs; cedar needles on the roof had gotten into the elastic seam of my shorts.

Marea, Lassie and the bluestone tanks used in cleaning algae from fishing nets. Photo courtesy of Marea Larsen.

THE GOOD HOPE CANNERY

There was little time for leisure once the fishing season started, but on May 24 the Good Hope crew enjoyed a ride on the *Ermalina* to Clam Beach for a Victoria Day picnic. Photo courtesy of Marea Larsen.

MAREA'S STORY

Seated left to right: Harold Wood, Roy Chaplow, Agnetta Gulbranson (Ludvig's wife), Marea Abelson (back turned), Irene Abelson (Olav's wife), Alma Gulbranson, Jake Skrindo, Charlie Gulbranson (Ludvig's uncle), unidentified Ermalina crew member, Pete Pederson, net boss Ludvig Gulbranson. Standing left to right: unidentified Ermalina crew member, Olav Abelson, Jergen Aagaard, Ken (last name unknown; storekeeper), Hans Peterson, Hans Christianson.

THE GOOD HOPE CANNERY

The Japanese were well represented in Rivers Inlet prior to 1942. With the attack on Pearl Harbour came confiscation of their boats and internment far from the coast. Toshichi Miki, wife Niimi and their children were friends of the Abelsons. Marea's playmate Mary is holding a doll. Photo courtesy of Marea Larsen, c. 1940.

One night a cougar killed all our cats except one named Me-Oh-My. I remember hearing the big thud of the cougar's paws jumping on the outside stairs of the cookhouse. Eventually Dad killed it with some wolf poison. Poor old Jiggs was also killed by a cougar.

In the fall Dad would salt fish for feeding the cats in the winter. Any bread that didn't sell he would put in the unused bedroom above the kitchen in the cookhouse, and this bread would dry out and became Melba toast, which was added to the salt fish to make cat food.

We had a parrot that had once belonged to Mr. Sandell. [A Good Hope veteran after whom the lake in the mountains behind Good Hope and the creek flowing from it are named.] He married an Indian woman and she died quite early. In the winter he liked to go up and visit her family up near Bella Bella or Bella Coola, so he started asking if we would keep Polly, which he had bought for his wife, so we looked after the bird.

On the May 24th holiday weekend we'd go to "Clam Beach" by packer boat for a picnic. I wasn't allowed to play with the Indian children because my parents were afraid that I might get TB from them. But there was never any worry about the Japanese children. Mary Miki and I were such good friends. Mary liked to stay with us when her mother and father went out fishing. Mom used to refer to us when she tucked us in at night as "My little white-haired girl and my little black-haired girl." We played together and once we were trying to put clothespins on our noses and she couldn't do it, it wouldn't stay on her nose. She asked me, "Why does it stay on your nose and not on mine?"

There were enough Japanese fishing at Good Hope before the war that there had been a bunkhouse for them, but there were never as many of them as there were Indians. On December 7, 1941, Dad and Mr. Miki were listening to the battery-powered radio at Good Hope, ears glued to the set, listening to news of the bombing of Pearl Harbor. Mr. Miki exclaimed, "Big war now!" Two days later he and his family were gone, taken away by Canadian government authorities. After the war there were a few who came back, but not in the same numbers. The Japanese bunkhouse had a nice big bath. I don't think anyone ever really lived in the Japanese bunkhouse after the war.

One of my jobs was to carry our drinking water back to our house from a tap near the Indian shacks. The water came from the lake. The main criteria for a cannery location was a steady supply of drinking water. They had well water, but it was yellow cedar water, so we packed our drinking water. Even though I lived on or near the water I never learned how to swim. I didn't go out fishing with my father because the one time that I tried it I found it boring because at night I slept while Dad put out the net, and he slept during the day.

Down from our house was a house on the beach that was built by a man named Erickson. He and his family were gone before we ever got there. He'd had an Indian wife and quite a large family and it was a very nice house, but it was not used by anybody, so we used it as our chicken house. So there was a grand life for our chickens because there was a front room

THE GOOD HOPE CANNERY

The binders on the shelves in this photo of the Good Hope office may well be the ones found stashed in the wall in 2006. Photo c.1940s. Left to right: Ludwig Gulbranson, one of two Good Hope net bosses; Harold Wood, bookkeeper; Levi Lauritsen, manager. Photo courtesy of Mildred Dalton.

where their food and water was and a second room where all the nests were and their roost. They were happy chickens and they had nice windows to look out of, but in the yard there was a pretty little creek and they had fresh water to drink outside. One year we had five roosters, and so I named them Roosevelt, Mackenzie King, Churchill, King Haakon after the Norwegian King at the time, and Stalin.

Like most girls, I enjoyed playing with dolls, and I had a nice collection. The Erickson family had many young children, but one had died at birth and they buried her in the yard under one of the currant bushes. There was this little wee cross to mark the spot and one day—I was six or seven—I was digging where the cross was. And Mom, who was a very understanding person, found me and said, "You know, it wouldn't be like a doll." She had enough presence of mind to realize what I, perhaps as a somewhat lonely little child, was thinking. Of course she pointed out that there'd only be a skeleton. In retrospect I think I was a little lonely for the company of other children, but what worried me the most was that when I went to a real school that people might think I was different.

[Good Hope's manager,] Mr. Lauritsen, was very proper. He wore a business suit and tie at the cannery, and he ran a tight ship, probably a lot tighter than others. He treated all of the employees at Good Hope with strictness, and you wouldn't call him Levi to his face, it was Mr. Lauritsen and Mrs. Lauritsen, there was that certain decorum. He built a building at the curve of the bay to use as a social place, but no one ever used it, it was a real white elephant, it just sat there. He arranged one or two card parties in the cookhouse sitting room when the crew was getting ready for the fishing season. The Lauritsens would provide coffee and sandwiches. Everybody enjoyed it, but they came mostly because Mr. Lauritsen said so. He seemed to never get along with the cannery's cooks. One day the cook took off his apron and caught the next boat out. Mr. Lauritsen went through two cooks that season. Then he asked my mother if she'd take over the cooking duties, which she did. But when he asked her to do it again the next season, she refused. Well, you didn't have words with Mr. Lauritsen! He was quite cross with my mother, and so was Mrs. Lauritsen, who didn't speak to my mother for a long time. Coastal people could be quite eccentric, they had "different" ways of acting, but people were more tolerant in those days; it was "'live and let live." More so than today. There was quite a cast of characters at Good Hope.

Hawaiian-born Maggie Johnson was a close friend of the Abelsons. She and her Norwegian-born husband, Fred, lived near Goose Bay and were frequent visitors to Good Hope. Photo courtesy of Marea Larsen, 1940s.

Maggie Johnson was of Hawaiian descent. She and husband Fred, a Norwegian-born fisherman, lived year-round near Goose Bay. My mother had a straw hat the mice had gotten into, but that didn't bother Maggie; to her it was still a nice hat to wear.

THE GOOD HOPE CANNERY

Mr. Glendinning was the watchman and somebody asked him why he'd never married and he said, well, the ones that he was interested in weren't interested in him.

John Larsen we used to call "Smokestack John" because he was very tall and always very grubby looking. One time on a visit to the hospital a nurse recommended that he change his underwear, but he couldn't see any reason to. He wore the same clothes for months on end.

Mr. Ovitslund and his wife had three children. Their eldest was son Thor, Margaret was their daughter, and they had a younger son named Norman. At Thanksgiving in 1947 Thor and Margaret had been to a dance in Aldergrove and were walking along when a drunken driver came and killed her instantly. The next May 24th holiday Thor was working at Good Hope with his dad. He went up to Sandell Lake behind the cannery with two others to do some fishing. He went into the canoe that was up there with his cork boots on and the canoe sprung a leak. The water was frigid cold and he drowned. So here was poor Mr. Ovitslund and his wife left with only one out of three children within one year.

Oscar Soderman was a teacher in the winter and he always came up every summer. He had this tiny little boat although he was quite a big man, tall, and well formed. Some people used to be sensitive that he fished because he taught and was fairly well paid. There were those who thought he was "double-dipping."

Albert Gjertson was the net boss at Good Hope, but when McTavish Cannery closed ABC didn't want to lose Ludvig Gulbranson, who was also a net boss. Both men had quite a following. A fisherman would have a net boss that he liked and even though it was the same company the fishermen would go to that net boss to get his nets. Fishermen had their loyalties, like going to a garage to a certain mechanic. One mechanic does the same work as another but some are better. So when McTavish closed, instead of letting Ludvig go, they did something that no other cannery did, they had two net bosses. They created one net loft for Ludvig and one for Albert. I often wondered if there was a certain amount of animosity between the two. Two cooks but they had separate kitchens. Ludvig Gulbranson was also a top-notch fisherman. He studied the moon and the tides, giving him a much better idea where the fish would be. Ludvig made a real science of it. Rumour was that Lauritsen was jealous of Gulbranson because he was making as much money fishing as he was as net boss. So eventually Ludvig left Good Hope.

Pete Pederson was the net man. He used to be a fisherman, but he lost a leg in a fishing accident. He lived with Paul Gartner. Gartner and my Aunt Dell had a little romance going.

The cookhouse, 1940s. The kitchen and dining area were on the main floor and there were bedrooms on the second floor. Photo courtesy of Marea Larsen.

Rosemary and Gerry Miller with newborn Ken and Nancy Anderson, 1953. Photo courtesy of Gerry & Rosemary Miller.

Mr. Wood was the bookkeeper. He was a great storyteller. None of them were true, but he was a good storyteller.

We lived in different houses over the winters; it was our choice of living quarters. We stayed for a number of winters in the cookhouse building. My bedroom was upstairs, overlooking the bay. There was a crack in the window so I had fresh air coming in all winter along with Toby, a cat that the Indians had left behind one season. He would crawl through the crack in the window and spend the night with me. One day my mother asked about some little spots of blood on my bed sheets. She wondered if I had a cut somewhere, but no I didn't. It was Toby's paws; he'd nicked them on the cracked glass on his way in.

In the winter we would cut ice off the tops of the empty, overturned bluestone tanks and use it to make ice cream. It made the best ice cream. We used Pacific condensed milk, which was like using cream.

By the late '40s my parents were ready for a change and Mr. Lauritsen took it upon himself to keep an eye out for a new home for us. He was going to make sure that we got the right kind of place when we moved to the Lower Mainland. He called up Mom and Dad and said that he'd found a place for us. It was too big for he and Mabel, but it would be good for us. And so they took Mom and Dad out to have a look at it and they bought it. Mom was always looking forward to living in the Lower Mainland. We moved to the new house in Maple Ridge in November of 1950.

I graduated from high school in 1952. That summer I worked at the *Gazette* newspaper, but in the summers of 1953 and '54 I worked in the Good Hope office as an assistant bookkeeper. Unfortunately, Mom didn't have many years to enjoy living in Maple Ridge before she died of cancer in 1956. Through that time the Lauritsens were really good about helping. The day that she died I called the Lauritsens and they came and stayed until the funeral people arrived, so that was the kind of friendship we had. Mrs. Lauritsen helped with our wedding, setting up tables and things. The Lauritsens were excellent neighbours, though they didn't live close by. We were good friends and I've stayed in touch with other good friends from Good Hope: Gerry and Rosemary Miller. Gerry was the mechanic at one time and Rosemary was the store assistant. And I'm in touch with Mildred Dalton, she was Mildred Lauritsen, Levi's daughter, and with Inga Fenwick, she was the storekeeper for a couple of seasons.

Gerry Miller and his son Ken as a toddler, already at home on the docks. Photo courtesy of Gerry & Rosemary Miller.

Marea Abelson and Astrid Gjertson, 1950. Photo courtesy of Derek Low.

Chapter 6

Square Hooks

I don't believe in catching one fish at a time, just gives the bastards a chance.
—A commercial fisherman

In July 2008 I took my first trip to Good Hope. It coincided with a group of ex-cannery managers and fishermen who had been invited up, including Ken Mack Campbell, author of *Cannery Village: Company Town*, a history of British Columbia's outlying salmon canneries. The informal group, billing themselves as "The Cannerymen's Association," normally met once a year for lunch, so a trip to Good Hope was something special. I knew one of the group, Dick Nelson, a former president of BC Packers and the eldest son of West Coast legend Richie Nelson, the driving force behind Nelson Brothers Fisheries, at one time the second-largest fish packing company on the coast. I had hoped that one or two of the group might have worked for ABC in a management capacity, but, alas, none had. Nelson Brothers, Canfisco, BC Packers, Millerd and Gossage, and a variety of other fishery businesses were represented, but no managers or workers from Good Hope's original owner.

There was, however, one Good Hope original resident present. Marea Larsen had come up, her first time back since the summer of 1954. She was accompanied by her husband, Terry, a first-time visitor. If Marea was excited or anxious about the trip, she didn't show it. The soft-spoken 73-year-old, looking more like 50, was calm and composed. And why not? Who knew better than she exactly where she was going?

An hour after the float plane docked, Marea, Terry, Tony Allard and I toured the

THE GOOD HOPE CANNERY

The Abelson cottage overlooking Good Hope cove, 1940s. Photo courtesy of Marea Larsen.

manager's house. Built by Levi Lauritsen in 1940, in part from materials salvaged from the McTavish Cannery, the house was still in use, interim manager Doug Montgomery bunking on the premises. After confirmation from Marea that the house was, indeed, as unchanged as it appeared, we moved on to the larger cookhouse, just steps to the southwest. This two-storey structure had served in the Lauritsen era as a kitchen/dining hall on the main floor and as sleeping quarters on the second floor. It was now in rough condition: floors sagged, walls leaned and everywhere were signs of mildew and rot. The building was used for storage, which added to the sense of dilapidation. We made our way through the rabbit's warren of rooms, stepping carefully over and around piles of junk, Marea pointing out what each room had once been used for. We climbed a set of stairs leading to the second floor. At the top we turned left and immediately left again into a small room. This had once been Marea's winter bedroom. A window offered a view of the cove and the three islands. The walls held four or five framed pictures of cats, evidently cut from 1970s-vintage magazines. There was nothing else on the walls, just the cats. Obviously a cat lover had once occupied the room. Even so, it was a bit odd. And then I remembered Marea's cats; she'd had half a dozen at one time. But hadn't she mentioned one in particular?

"Marea, what was the name of that cat of yours?" I asked. "The one that cut his paw on your broken bedroom window pane?"

"Toby."

"These aren't your cat pictures by any chance?"

"No, they're not mine," she said, laughing. "These are well after my time."

"I didn't think so. But it's kind of odd, don't you think, that there'd be all these cat pictures in this room and not anywhere else in the building?"

The Manager's house was strategically placed, with a commanding view of the cannery, wharf and water. Photo courtesy of Marea Larsen.

"Yes, it's a little odd, I guess!"

"It's spooky."

"Yes, I can see that. Who knows, maybe it's the ghost of Toby come back to haunt us," she joked.

One afternoon I spoke with Ted Walkus, a fishing guide in Rivers Inlet since 1970. Born in 1957 in Bella Bella, Ted believes he may have been the last child delivered by Dr. George Darby and recalls vividly carrying the cross to lead the procession at Darby's funeral in September 1962. I asked him what the cove harbouring Good Hope had meant to his people. "The site of Good Hope Cannery is not on a salmon river, so if you were to rate its historical importance to our people it would rate a one out of ten, with ten being very important. It may have been a source of crab, eelgrass and spawning herring. Certainly it was a safe place to harbour in bad weather, so it has some importance that way." What could he tell me about the Oowekeeno in relation to Good Hope? "Wadham's and RIC were the domain of our people." [Rivers Inlet Cannery was the first cannery in the inlet and, being at the mouth of the Oowekeeno River, was nearest to their village.]

Ted's grandfather Simon provided his stories for a 1982 book, *Oowekeeno Oral Traditions as Told by the Late Chief Simon Walkus Sr.* Simon Walkus Sr. was a high-ranking chief of the Rivers Inlet people and one of the few remaining elders well versed in Oowekeeno culture as it existed at the turn of the twentieth century. The book, transcribed and translated by his daughter, Evelyn Walkus Windsor, contains twelve stories and three songs. "The Boy Who Turned Into a Salmon" tells the story of the spring salmon reproductive cycle. "In Oowekeeno cosmology," writes Walkus Windsor, "the spring salmon (also known as tyee, chinook, or king salmon) is the highest ranking of all salmon and is therefore entitled to a ritual sprinkling of feather down, an honour reserved for chiefs."

> It is said that a child was lost in an abandoned village. In vain people searched, after which his father and mother never left this place and just lived there. After four years this child returned. The fish trap of the parents had been finished and the child entered it. To his father's surprise

his child was sitting inside it. So he pulled it up. It was his child to be sure. His child went to the house and began to advise his parents on what they were to do. "Make a new fish cutting mat and spread it at the doorway for us to lie on when you catch us in the morning," he said. "Put feather down on us. Our right hand side shall be the side on which we lie when you bring us inside the house, and you shall put feather down on us so that we are adorned with feather down." Having been adorned with feather down for a while when entering, the fish arose. Then after staying there overnight, he made his parents young again. His parents became very young when he made them be young again. "Well then, we shall leave you tomorrow. Two years from now I shall come to see you," the child said to his father, who left to put him in the water when it became day. The child went up the lake to the head of Owikeno Lake. The fish began to dance, timing the length of their stay, and returned, left for the place the fish go to in the ocean. After two years the child returned again. He had six children by then. This time also they entered his father's house and were fed. They also only stayed there overnight. The father put them into the water and they began to swim away to the dancing in the lake. They returned again and straight away went to the ocean, the destination of the spring salmon growing up.

This story reminded me of Chief George's account from *Living On The Edge*. As a boy he had travelled with his father a long distance over the sea until he arrived at Good Hope. What was Good Hope for most of the year but an "abandoned village"? While his father fished, he worked filling baskets full of cans, the equivalent of the basket-shaped fish trap in the Oowekeeno story. After all, cans can be "caught" in a basket as well as fish. The story of "The Boy Who Turned Into a Salmon" was also Chief George's story. The symmetry was complete except for the part about making his parents "young again." But it occurred to me that Chief George had, in a sense, made his father "young again." When a father shows his son "the ropes," doesn't he, in effect, make himself young again?

The only man at Good Hope during my visit in 2008 with a direct connection to the cannery was a "high liner" by the name of Karliner. Bob Karliner was a fit and feisty 88-year-old of German heritage who had started his fishing career on the Fraser River in the 1930s. We had a number of short conversations over the course of four days. The first came one evening before dinner. I had noticed that he hadn't been out fishing since we'd arrived. "Not interested in fishing, Bob?" I asked, realizing as I said it how stupid it must have sounded given his seven decades as an outstanding fisherman.

"No, I'm afraid I won't be, not this time. But that's okay, I've caught my fair share."

"Did you ever fish in Rivers Inlet?"

"Oh yes, I sure did."

Some of the Good Hope fleet anchored at the dock. Note the wooden bluestone tanks lined up on the wharf. c.1948. Photo courtesy of Rolf Hundvik.

"Out of Good Hope?" I asked, hopefully.

"Yes, as a matter of fact, I delivered my first fish to this dock in 1937."

Jackpot! With any luck, Bob would have stories and know them all, Gulbranson, Wood, Lauritsen. "Levi Lauritsen was the manager then," I said, hoping to engage his memory.

"Lauritsen? Yes, I remember him."

"What can you tell me about him?"

"It was a long time ago."

"Anything about Good Hope stick in your mind?"

"Well, no, not really."

As he talked about his life as a fisherman, I could tell that he wasn't thinking about Good Hope then and he wasn't now. Good Hope had been just another cannery, a place to sell his catch. For him the action wasn't at the cannery, it was out on the sea. As he talked it was as if a switch had been thrown somewhere deep inside of him. There was excitement in his voice as in his mind's eye he saw boats, crews and catches anew. He couldn't get out the stories fast enough. "I've just published a book," he said. "It's all in there."

When I got my hands on a copy of *Stand By—Let 'er Go!* I read it with great interest, including his account of going up to Rivers Inlet as a 16-year-old fisherman in 1937:

> We left for Rivers in *Lillias*, travelling up via the Yaculta Rapids to join a large fleet off Pulteney Point at the north end of Malcolm Island. We soon found out why. They were all waiting for the weather to improve before crossing Queen Charlotte Sound like a giant fleet bound for Rivers and other up coast points. One fellow there was a Norwegian called Nelsen, who was good enough to take the time to explain everything we should know about working in Rivers. We found generally, as newcomers, that the older more experienced men were usually very helpful with good advice and proper explanations. We were still children and maybe they took pity on us, but I guess you could also call it the camaraderie of the sea—and it sure helped.

He continues, describing the terms under which he fished for ABC, which means that he's describing either McTavish or Good Hope.

THE GOOD HOPE CANNERY

We finished up the season at Rivers Inlet by the first week in August. We had to pay rent for our net and buy insurance for shark damage. All the canneries issued their own coins which was what we used for purchasing supplies from that same cannery. It was a form of bondage by which the cannery ensured the loyalty of its fishermen. Effectively, they owned you and, by the time we got back to Vancouver and everything had been settled up with ABC, I had $60 left for our five weeks of work, being away and roughing it in all conditions.

Built in Rivers Inlet in 1918 by Bell-Irving's contemporary R.G. McTavish, McTavish Cannery was sold to ABC in 1932 and shut down permanently in 1939. Photo courtesy of Marea Larsen.

116

An observation in *The Fisherman* newspaper in November 1939 extends Karliner's recollection. "In 1909 if you had $100 on September 1, you could live on it till Xmas if necessary. Now if you have $100 on September 1, you can live on it till September 10, if you are careful."

In his book, Karliner speculates on what a cannery manager looked for in a fisherman. While he could be thinking of any number of managers, he is looking back to 1937 and that means Lauritsen or managers of his ilk.

> I guess that was one of the skills of the cannery manager to spot a youngster with good potential and bring him along. If he knew you were a worker and could be relied on to produce, those were attributes that he took note of. But if you managed your affairs in a clean and tidy manner, had a good deal of common sense and knew how and when to take a measured risk then he had a winner and I would have to say with some modesty that I believe I measure up in all these departments.

Manage your affairs in a clean and tidy manner… a good deal of common sense… a measured risk… Levi Lauritsen would have agreed.

~

Morning comes early at Good Hope Cannery. There was a sharp rap on the door at 5:00 a.m. "Good morning, this is your wake-up call." And so on down the hall. I swung my legs onto the floor, Pat "Salty" Simmons' parting words after the previous night's dinner ringing in my ears, "You can't catch fish in your pajamas." It was one of those clichés that you're doomed to remember for the rest of your life. An hour or two later I watched as Chris "Hans" Sprathoff filed a barb-less hook, the only salmon fishing hooks allowed by regulation. "We like them to be what we call 'sticky sharp,'" he explained. "Feel that." I ran a finger gingerly over the pointed end. It was indeed "sticky sharp." By Rivers Inlet standards, the fishing had been poor over the last few days and the guides did their best to keep up everyone's spirits. As we trolled an area on the south side of the mouth of the inlet called "the wall," one of the other

THE GOOD HOPE CANNERY

Wadham's, south of Good Hope, was built in 1897 and was the largest cannery operation in Rivers Inlet. Photo courtesy of Marea Larsen.

Good Hope guides came over the radio. "How's it going, Hansy?" Hans picked up the radio, smiled sardonically, and replied, "We're not catching anything, but we're washing some nice herring."

An hour later, with still no action, Hans got on the radio and said quietly to Ted Walkus. "What's happening, Ted?"

"Nothing, Hansy. Maybe we need to try square hooks."

Hans chuckled. "Roger that."

"What does he mean by 'square hooks'?" I asked.

"He's joking. Square hooks are nets." I remembered what one of the cannery men had said to me the previous day when I'd asked him how he liked sport fishing compared to the commercial variety: "I don't believe in catching one fish at a time, just gives the bastards a chance."

As we were heading back to the cannery later that morning, Hans pointed out two small, plump, white-and-brown-mottled birds skimming along the water at a high rate of speed. "Those are marbled murrelets," he said. "They're an endangered bird."

The marbled murrelet is a seabird belonging to the auk family. Well adapted for life on and below the sea, the murrelet's feathers are unusually thick and dense to keep out the cold. Its feet are webbed like a duck's, but are small and only used for propulsion when it swims on the surface. Underwater it uses its feet to steer and is propelled forward by its powerful wings. Its bill is well designed for seizing slippery fish with a firm grip. The murrelet breeds in the summer along the BC coast and is dependent on old-growth forests for nesting. It spends most of its time fishing, diving well below the surface and swallowing its prey—herring and sand lance—underwater. It is believed that a murrelet can dive to a depth of 140 feet (42 metres), possibly deeper. It relies on dives lasting less than thirty seconds. A murrelet will sometimes drive a dense school of fish to the surface and keep it there as long as possible with shallow dives. During these dives the murrelet will pick off and eat individual fish. Often, after a series of dives, the bird will flap its wings vigorously, fluffing its plumage and restoring the insulating effect of its feathers. A brief splashing bath usually marks a successful fishing effort. The swarms of small fish boiling at the surface attract other murrelets and gulls eager for easy pickings. The murrelet adapts its dives to the fish available. In the south, sand lance are attacked in small channels and inlets where they are concentrated by tidal currents. In the north, herring seem to be caught most easily against steep cliffs in deep inlets. Murrelets also feed on northern anchovies. In May and June, small fish are so widespread that the birds can fish almost anywhere and spend much of the day preening and dozing. In the summer, fish become more scarce and murrelets concentrate their efforts on richer areas such as tidal rapids. If fish are unusually plentiful more than one hundred birds may gather in a flock. The murrelet will carry a meal as far as 40 miles (64 kilometres) from the sea to its nestling every night for a month.

The Union Steamship Cardena, a vital link to the world, unloads at Good Hope, early '50s. Photo courtesy of Gerry and Rosemary Miller.

Letters Home

On March 31, 1927, a 22-year-old Dutchman, Adriann Jacob Sem Van Alphen, boarded the SS *Camosun* in Victoria bound for a job at J.H. Todd & Sons' Beaver Cannery in Rivers Inlet. In a series of letters he wrote to family back home in Holland, he describes loneliness and its symptoms.

"Once a week we get mail here," he writes, "and at that time the whole cannery is a bit disorganized, it is really funny the influence mail has on the mind of the people who work in a lonely place, far away from their relatives and friends. He who has mail is happy and the one who has no mail that day is sincerely pitied and consoled by those who were fortunate to get some."

There is also boredom to contend with and the pain everyone seems to have suffered in being thrown together in a confined space with nothing to do except work. A story he relates about the cannery manager's dog could just as easily apply to the cannery's two-legged creatures.

"We have a dog here belonging to the manager, an Airedale terrier. He just performed a feat that all of us, men, are jealous of. One day he was walking around the docks with his nose in the air, sniffing smells that interest dogs, and just as we all expected it was from a female dog somewhere in the Inlet. Finally Mr. Murphy, that's his name, jumped in the water, he just could not stand the temptation any longer and swam out of the bay and into the Inlet proper, the water was not just the nicest for comfortable swimming to our liking, but hormones do stimulate metabolism tremendously of course, and [the dog] swam to the next Cannery some 2 miles from here… and did not find her waiting for him, heard that later from the crew there. That little mishap did not deter Murphy of course, he jumped into the chuck again and continued his search and a mile or so further up the Inlet his quest for matrimonial adventures was rewarded and he did what was expected by Mother Nature, made dead certain that the dogs were saved from extinction as far as Murphy was concerned.

He was, as a matter of course back home on time for a good meal and the great admiration of the whole cannery crew." Van Alphen had little good to say about the Japanese and Chinese workers. He wasn't any keener on the British contingent, whom he had "learned to dislike and even hate." A British co-worker seems to be the source of his disgust.

"The lineman, an Englishman, starts to whistle and to sing as soon as he opens his eyes in the morning, all the time and never stops, during his washing and shaving, comes to dinner and other meals doing the same, all the time he is working, during the passing of the dishes when eating, comes home late in the evening, and still going strong, sings like a fool and we all become fools listening to his incessant singing and whistling, does not stop for one minute it seems, never talks, except when he knows that everybody is sleeping or trying to sleep, then he talks loud till about 12 or 1 in the night. His name… strangely enough: Jimmy Friend."

Only one nationality wins his approval. "I have learned to like the Scotch, for some reason or another I feel close to them, awfully nice people they are." A Scotsman named Leo MacKenzie seems to have been largely responsible for the warm feelings. "The Net-boss is the son of a very wealthy Scotch family, who became impoverished by a bad deal in some kind of trade. He is a dyed in the wool 'gentleman.' He is very fond of classic music and has a collection of discs and a gramophone here with him. He loves to play them and drink whiskey. I think that the whiskey may be a bit of a weak point in his make-up."

Another co-worker wins his sympathy: "The bathroom man is a former deep sea diver, badly crippled by 'the bends,' the result of diving for the salvage of the Quebec Bridge's centre section."

Genuinely impressed with the Natives, Van Alphen writes: "The Indians are well educated

THE GOOD HOPE CANNERY

here and most of them speak good English. They are not stingy with their money, they spend it as fast as they can. Good eats, boots, clothes, nice things for the women etc. They are a very gay and easy living people, seldom is a nicer and fuller laugh heard."

Some years later, in a postscript, he recalls people and events omitted from his letters. He was fascinated with the Natives' gambling game, *slahal.* "[The] gambling element consisted of one or the two opposing parties, in turn, juggling several little sticks behind the back while standing up and as soon as the singing and tricking stopped they, the opposites had to guess in which hand the sticks were. All the time the defending party had to draw the attention away from the juggling of the person standing. And that was for them the fun. What they gambled for I never knew. But there was one woman, Loessie by name, when the tension was at its height, suddenly got up and loosened her blouse and start[ed] to juggle her breasts and that did the trick, all laughing so much that her party won out."

Van Alphen talks of sexual favours extended, wild escapades and a boatload of prostitutes travelling the route of the camps and canneries. At each stop, they gave the manager the honour of free services; the rest had to pay.

These "free services" did not make Beaver Cannery's manager any more likeable. "One bad day," Van Alphen writes, "the Cannery did not want to pay the price for the piecework of the Indian women and they really worked hard and good, they just said… we do not want to work for less, we are not even asking for more money. They did not go on strike, they just walked out, without saying a word to anyone, they packed up and even left the Cannery in a few boats, going I do not know where, the men, the fishermen, left also. We do not need you white men, they quietly said, we lived long before you came to this country. And for the rest of the season we had to try to do what we could ourselves, since not one Indian woman came, even for more money, the solidarity and the dignity won out over the Cannery Management."

That night over dinner I spoke with Orville Otteson. His father, Hans, managed Wadham's Cannery in the 1950s and '60s and Otteson and his brother spent many summers there. It was evident from the tone of his voice that those summers had been a golden age. "It was a great place for a kid to be, there was always something going on. One time we all had our hair cut Mohawk style. When the *Cardena* came in to dock we'd shoot arrows at it as if we were Indians on the attack. Playing Cowboys and Indians was a big thing back then. They'd play

The Union Steamship *Cardena* docking at Good Hope. Photo courtesy of Marea Larsen.

John Ford-style westerns on Saturday nights. The place would be packed. When the Indians on screen were massacring the cowboys the Native guys would cheer for them, and when the cowboys were winning the white guys would cheer. It was all in good fun. We'd watch the Natives playing stick games and we'd go on picnics out to Clam Beach. Back in those days there were so many boats in the inlet, hundreds and hundreds, you couldn't get from one side of the inlet to the other."

One afternoon on our way to fish for chinook at the head of the inlet, Cyril Douglas took us to see the "dancing man," a Native pictograph located on the eastern side of the entrance to Moses Inlet. Painted on an inward slanted rock face about 10 feet (3.04 metres) above the high-water level, the main figure appears to be a fierce man or creature with an open mouth, his arms spread wide as if he is running wildly at the viewer. To his left is a long, serpentine squiggle topped with two or three sets of marks, one of which resembles the scratches left by a bear's claw. Douglas didn't know what it represented and neither, when I asked him later, did Ted Walkus. No one, it turned out, including First Nations people, knows with any degree of certainty what pictographs such as the dancing man are meant to communicate. Their meanings are lost to time. We didn't linger too long at the pictograph. Salmon were waiting for us, we hoped, in the waters just off a place known as Pigeon Point. As we waited for a strike, I noticed two or three of Pigeon Point's band-tailed pigeons flitting from tree to tree.

The band-tailed is the biggest pigeon in North America, soft grey like the rock dove, but longer and sleeker. Its bill and feet are yellow and its eyes are black with a thin, red, fleshy eye ring. Adults have a white collar at the nape of the neck with an iridescent green patch beneath it. Their breasts and bellies are washed in a pinkish-mauve. Band-tails are found in low- and mid-elevation forests. Coniferous and mixed coniferous/deciduous habitats are both used, as long as there are some large conifers in which to roost. Band-tailed pigeons prefer forest edges—open sites bordered by tall conifers. In the early breeding season, mineral springs and tidal flats become important. Toward the end of the breeding season, many birds migrate to higher elevations to feed on ripening fruits. They are sociable, foraging in flocks most seasons, and nesting in small colonies. Most foraging takes place in trees, where they climb around with ease, although they also feed on the ground. Their diet consists of berries, acorns, seeds and pine nuts. In June, and again in the late summer, they congregate at mineral springs where they ingest salts. In late summer they also feed more heavily on fruit. Loose colonies with several pairs nesting together are common. Nests are usually located in trees, 15 to 40 feet (4.5 to 12 metres) off the ground and are typically placed in a fork on a horizontal branch, or at the base of a branch against the trunk. The nest is a loose, bulky platform of twigs that the male collects and the female puts in place. The female lays one or two eggs, and both parents help incubate them for eighteen to twenty days. Both parents feed the young "pigeon milk," a protein- and fat-rich liquid produced in their crops. The young stay in the nest for about four weeks and are tended by their parents for some time after they leave the nest.

Looking at a map of the band-tailed pigeon's breeding grounds, I was struck by its habitat overlaying that of the west coast fishery. Band-tails and fishermen have been occupying the same areas in the same seasons for a very long time. The ancestors of the Pigeon Point band-tails that flew above us that July day once flew above Oowekeeno fishermen and later above the frenzied manoeuvres of the gillnetters. Their treatment by human beings was identical to that of the salmon. In *The Birds of California*, Leon Dawson describes hunters converging on Santa Barbara County's valleys in 1911–12 and shooting half of an estimated half a million band-tails on their wintering grounds. "A humiliating example of what human cupidity, callousness, and ignorance, when unrestrained, will accomplish toward the destruction of birds." It brought to mind the staggering numbers of salmon slaughtered in the same era in Rivers Inlet.

We fished for three hours that afternoon. Not a single bite.

Late one afternoon I set out to see what remained of the little house that had once stood on the rise at the end of the cove, Olav, Irene and Marea Abelson's home. From 1895 to 1969 I wouldn't have had to scramble over sharp, slippery, rockweed-covered rocks to reach the end of the cove; I could have taken a leisurely stroll along the boardwalk. But by 1969 the boardwalk had fallen into an irreversible state of disrepair and H.O. Bell-Irving's grandson, Ian Bell-Irving, having taken over the helm of H. Bell-Irving & Company in 1962, had what

THE GOOD HOPE CANNERY

was left of it destroyed—almost all of it. Against the odds, one section of about 4 feet remained, covered in moss, hidden under overhanging cedar branches. I continued to the beach where a stream emptied into the sea. The butt ends of a few old piles showed through the mud and the tangled, overlapping remains of a wooden boat littered the area. I made my way around the wrecked boat, pushing through patches of comb-like alders. My first stop was the dilapidated building that I had seen from the cannery. It was in rough shape, looking like something that had been uprooted and washed down a mountain in a heavy spring torrent.

Shag carpeting and cocktails: Good Hope was reborn as a sport fishing lodge in 1970. Henry Ogle Bell-Irving's grandson Ian is at centre in the middle row, hand on knee. Photo courtesy of Barb Quinn.

But evidently it had been built here, straddling the stream. To what purpose? Months later I would ask Marea Larsen and the Millers if they knew anything about it. Marea drew a blank, but Gerry thought it was probably the old Japanese bunkhouse. I opened its back door, certain that if I stepped inside the whole thing would come crashing down on me. Poking my head in, I counted four rooms, imagining them occupied by Japanese Canadian fishermen such as Bunkichi "Kats" Nasu, Haruo Terashita, Takeo and Takemi "Tom" Miyazaki, and Toshichi Miki.

Good Hope's Japanese Fishermen

The Japanese Canadian contingent was never very large at Good Hope, but there were a few, including Bunkichi "Kats" Nasu, Haruo Terashita, Takeo and Takemi "Tom" Miyazaki, Ken Haminishi and Toshichi Miki.

Miki had fished from a skiff for ABC out of McTavish, transferring over to Good Hope when McTavish closed. He had lived at Good Hope and at Dawson's Landing. He married Tokie Niimi, and they had eight children. Interned during the war, he returned to fishing in May 1950 with the Great West Cannery, retiring in 1960.

Nasu was born in 1884 in Ao, Wakayama, Japan. From 1906 to 1941 he fished Rivers Inlet for Wadham's and, following internment, from 1951 to 1968 he fished for Good Hope on *Audrey* and *Sea Biscuit*. He and wife Chiyo Iwasaki had seven children. His family pleaded with him to retire, and he did, reluctantly, at the age of 85. He died in 1978.

Terashita was born in 1926 and fished for Good Hope from 1950 to 1968. His boats were the *Glow Mac*, *Lady Bea* and *Amber V*. He died in 1993.

Takeo Miyazaki, born in 1904, fished BC waters for 40 years, including Rivers Inlet for ABC after World War II. His son, Takemi "Tom" Miyazaki, born in 1929, fished for Good Hope beginning in 1951. His boats were the *River Queen* and the *River Queen II*. Takeo passed away in 1996, and Tom retired in 1997.

I scanned the rooms for any obvious clues to confirm or refute its former occupants and use, but could see nothing. The stream gurgled softly beneath the house. A breeze rustled the alders. Otherwise there was silence and nothing is more silent than an abandoned bunkhouse. I was beginning to spook myself, so, out of respect or habit or both, I closed the door and set off up the rise in search of Marea's old domicile.

Somehow I overshot the obvious location and ended up in an area that had been logged ten or fifteen years earlier. You take your life in your hands when you try to hike through a west coast rain forest that has been logged off. It is as if a monster had been playing a game of pick-up sticks and, having released the sticks, had walked away from the mess. Stumps poked up through the slash and a dense canopy of salal covered every sudden 8-foot (2.43 metres) drop into ankle-busting darkness. After my third drop into darkness, I was starting to think that maybe Marea's house hadn't been quite so far back in the bush as this. I fought my way back in what I hoped was the approximate direction of the presumed location of the Abelson house. In Marea's day her mother had kept an extensive garden of fruit, vegetables and flowers. I hoped that I might stumble upon a vestige of it, but as I stumbled along, trying to weave my way through a thicket of alders, it seemed impossible that anything created by human hands had ever existed here. And then I saw a cluster of day lilies and a smattering of daisies. These weren't native species; these were civilized perennials. I had, I believed, found the remnants of Irene's garden. Encouraged, I scoured the area for other signs of human habitation. Ten minutes later I stepped on something hard. Bending down I scraped away a layer of dirt and moss. It was a piece of tongue and groove board, about 3 feet long and 4 inches wide. There were other pieces scattered around. More searching uncovered what appeared to be a section of boardwalk. Was this the boardwalk that I'd seen people standing on in Marea's photo collection?

One afternoon we were invited to the Goose Bay Cannery, about a half-hour south of Good Hope. We tied up at the weathered old float dock and went up the gangway. Goose Bay, a Canadian Fish Company cannery, had ceased canning in 1957, the last to call it quits in the inlet. Some years earlier a group of nineteen firefighters had purchased the property and were doing their best to maintain and enjoy it as a base for personal summer holidays and fishing excursions. Despite the firefighters' best efforts Goose Bay was doing what old canneries do best: it was slowly falling apart. Railings had disappeared and panes of glass were missing. Major repairs had been carried out, but there was always more to do. Still, for something built in 1926 Goose Bay appeared to have stood up extremely well. The large, cavernous cannery itself was in good shape, as were the many buildings built out over the mud flats on an extensive network of piles. Joined by hundreds of yards of boardwalks, these buildings housed offices, a machine shop, a generator, recreational rooms and employee housing. As I toured the site I couldn't shake the feeling that I was trespassing. Unlike Good Hope, where most of the stuff of the canning business had either disappeared or been tidied up, Goose Bay seemed to have been deserted in 1957, leaving everything exactly as it had been. In my notebook I recorded some of the things left behind:

Hand trucks, porcelain sinks, ovens, rusty bedsprings
Bedsteads, rusty heaters, pallets, saw blades
Pumps, levers, belt drives
Ropes neatly coiled and hanging on walls
A Towmotor forklift (red with white trim, steel mesh roof)
Detecto-brand weigh scales
Old boilers, spools of cable
Cooker's 1, 2 & 3 "Pressure, hold, skim" indicators for retorts
Canada Dry-brand pop wooden boxes
SALMON labelled cardboard boxes
Big sign: GO SLOW
Old toilets covered in fallen ceiling debris
Half-flat salmon boxes
Boat flags

A sign posted on the wall outside of a toilet—WHITE MEN ONLY—stopped me short, making me wonder if I was really in Canada. As I wandered around I was reminded of the book *The Stone Angel.* In it Margaret Laurence's character, 90-year-old Hagar Shipley, describes an abandoned cannery not unlike Goose Bay:

A place of remnants and oddities, this seems, more like the sea chest of some old and giant sailor than merely a cannery no one has used in years. The one enormous room has high and massive rafters like a barn. The planks in the floor are a greasy black, stained by years of dark oil and the blood of fishes. Fragments of rusted and unrecognizable machinery are strewn about haphazardly as though someone placed them there for a moment, meaning to return to them and never doing so. Oily hempen ropes lie like tired serpents, limp and uncoiled in corners. Wooden boxes, once stacked neatly, have been scattered and jumbled, but each one still clearly bears its legend of class—*Choice Quality Sockeye, Best Cohoe.* Festooned like sagging curtains across barrels or draped along the floor in sodden musty folds, the discarded fishing nets must have been left by the last fisherman to come here with his catch.

The Goose Bay Cannery, as seen from the deck of the *Cardena*, 1950s. Photo courtesy of Gerry and Rosemary Miller.

Not far along the boardwalk I came to a building housing a pool table, dartboard, chairs and a number of cardboard-boxed games on a table. A side door led to a toilet and shower stall, advertised as being FOR FISHERMEN ONLY. An adjoining smaller room housed desks, chairs, filing cabinets and a large safe. There were ledgers on the desks and the drawers were full of keys, requisition forms and slips of paper with the names of fishermen and their boats. A canvas mailbag hung from a hook, ready should the *Cardena* make a stop. A National cash register sat on a table by the door, some of its keys marked BEER, HI BALL, WINE, SUD, FISH. There were a couple of marine supply catalogues, each about twelve hundred pages, weighing about 10 pounds (4.535 kilograms) apiece, displaying every conceivable item of boat equipment. There were stacks of 3-by–3-inch cards with prices marked on them in bright red lettering: 5 cents, 2½ cents, 8 cents. The filing cabinets were full of papers: receipt books, personnel books, injury reports. On the floor was a 16 mm Victor projector, still in its case, still waiting for Saturday night to roll around. Another building housed a late 1950s Viking TV set and a Spilsbury & Hepburn radio telephone. I wondered if they still worked, but I thought of *The Twilight Zone* and decided that touching them, let alone fiddling with them, might not be in my best interests.

Over lunch someone told me that in Goose Bay's boom days the water in the bay ran red with salmon blood and the sky was white with seagulls. The water was clear today, if tainted by rusting debris. A large boiler rested in the mud just off the cannery dock. There weren't many seagulls, but there were dozens of barn swallows, swooping in and out of the cannery.

Good Hope manager Levi Lauritsen always dressed the part—suit, tie and Homburg. Seen here in the manager's house, with wife Mabel and daughter Mildred. Photo courtesy of Mildred Dalton.

Chapter 7

The Storekeepers

The Union Steamship captains were unanimous in their opinion that the Cardena *was the most graceful and best sea-boat of any line ever to sail the coastal waters of British Columbia.*
—Gerald Rushton, *Whistle Up The Inlet*

Henry Levi Lauritsen was born in 1889 in Minnesota and attended school to Grade 8. A self-taught carpenter, he followed his uncle, Victor Larson, to BC around 1910. (Larson is an English corruption of Lauritsen, a Norwegian name.) Levi built ships for the navy during the First World War and married Mabel in Minnesota in 1915; they had two daughters, first-born Elvira, and Mildred, born in 1928. Victor was the manager at the North Pacific Cannery on the Skeena River before becoming manager at Good Hope in 1921, replacing R.E. Carter. Victor gave Levi his start in the industry operating an Iron Chink. Levi took to the business and went to ABC's Glendale Cove Cannery (purchased by ABC in 1911) in Knight Inlet where he was the foreman for many years. Eventually Levi became the manager at the McTavish Cannery. In either 1935 or '36 Victor fell and broke his hip, ending his days as manager at Good Hope. Levi replaced him and managed both canneries until McTavish closed down in 1939.

I met with Levi Lauritsen's daughter, Mildred Dalton, at her home in Cloverdale in September 2008.

THE GOOD HOPE CANNERY

My father was a Minnesota farm boy at heart. He was never a fisherman; he was always a cannery operations person. He disliked boats because they were so slow. He much preferred to put his foot on a gas pedal and go fast. He admired and respected anyone who worked hard. He had no tolerance for slackers. He was in favour of the working man, but did not like union methods. He disliked the way Homer Stevens [United Fishermen and Allied Workers' Union organizer] signed up young high school workers at Good Hope in the mid-1940s, because he didn't explain to them that their union dues would eat up their wage gains. He was greatly disturbed by the seizure of Japanese fishing boats, homes and other assets in 1941–42. It saddened him when he had to buy some of their boats for the company. "All that hard work for nothing," he said. The Indians would get permission from my father to use one of the buildings on the wharf to play their gambling game, *slahal*. We were allowed to watch as they chanted and sung, shaking bone dice in their hats. In the off-season Dad built and rented houses. He read extensively, mainly history, and wrote a Lauritsen family history. He was a big believer in the importance of education and would quote portions of Longfellow's "Hiawatha" for fun. He played golf at the nine-hole course where Empire Stadium used to stand. When he retired they bought a few acres in Haney, now Maple Ridge, and built a barbecue and a pond on the grounds.

One of the best and the last of the old-school cannery managers, Henry Levi Lauritsen. Photo courtesy of Mildred Dalton.

Mabel Lauritsen arrived at McTavish with Elvira and Mildred in 1932. Mildred started working on Good Hope's canning line in 1940, the last season of canning. Her job was picking out the cans containing the reddest of the sockeye meat, the most valuable. Thereafter

Mildred worked mainly in the store as an assistant, initially with storekeeper Norman Scott. Through the mid-'40s Mildred assisted Inga Thompson. Levi was always on the lookout for staff. Mildred's boyfriend, Jim Dalton, had worked part-time at a Safeway grocery store. Levi wanted to hire him for the store at Good Hope, but first sought his daughter's permission. It was granted and Jim went on to work three seasons at Good Hope.

> My father had the Delco plant, the power generator, shut down at 10 p.m. every night. I know because Jim used to do it. It didn't come back on until six the next morning. In the early days at Good Hope everything came up to the store in bulk, but by my time most things were packaged. We were also able to sell fresh meat because we had ice. The fishermen came in to buy supplies starting on Friday nights at around six o'clock. There were as many as two hundred of them and the store was fairly small, so it got quite crowded and hectic. Jim Dalton had to put in what he called "horse stalls" to control the flow. There was a steady stream until Sunday at noon when they went back out fishing. Purchases were mentally calculated on a slip of paper. One day a Japanese boy came in and scratched his fingers against his open palm, saying "paper." Well, "paper" was what Natives used to whisper when they wanted toilet paper, but I knew he wasn't looking for that. So, thinking how clever I was, I got him sandpaper. Of course, all he wanted was writing paper.

Good Hope's store would be packed with fishermen come Friday nights, c. 1953. Photo courtesy of Gerry and Rosemary Miller.

Being the cook at Good Hope was a demanding job. A lot of cooks came and went, sometimes in the middle of the season. I remember one of them leaving in 1946 I think, I would have

THE GOOD HOPE CANNERY

been eighteen, so I became the cook for about a week. Swiss steak was one of the few things that I could generally make well. But you had to get the oven temperature just right or you'd burn things to a crisp. I only burned the Swiss steak once, so I did pretty well.

During the war the *Cardena* and the other Union Steamship Lines ships were escorted across Queen Charlotte Sound by navy corvettes. There was a sailor posted on board with a rifle. The ships were blacked out for the passage. It seems ridiculous now, but back then rumours of Japanese submarines were taken seriously. When the *Cardena* or [any] other ship arrived at Good Hope everyone went down to greet it, no matter what time of day or night it was. Salesmen would come off the ship and try to get orders for whatever they were selling from the storekeeper, who would be there checking off supplies as they were unloaded. I remember the time the *Cardena* accidentally rammed the wharf at Good Hope. It's written about in *Whistle Up The Inlet* but the writer has the wrong cannery, unless it happened more than once. It backed up and then came straight for the wharf. There were lots of people standing there, but when they saw it coming at them, they started to run. I remember Cecil Fisher, the mechanic, running but realizing he had nowhere to go and just bracing himself. The collision was like an earthquake. Fortunately no one was injured.

The arrival of the *Cardena* was always a big event. When ships arrived everyone would go down to the dock to greet them, day or night. c. 1955. Photo courtesy of Gerry & Rosemary Miller.

The *Cardena* would come up frequently in conversation with people who remembered the war years, almost as if it were an old friend. Part of the Union Steamship Company's fleet that served the BC coast for seventy years, the 235-foot-long (71-metre) *Cardena* was described by

Gerald Rushton in *Whistle Up The Inlet* as having "132 cabin berths, with deck berths for sixty on her maindeck and had a license for 250 passengers. With a top speed of over 14 knots, (she averaged around 13 on the route) she was well adapted to the trade. She had two hatches and could carry 350 tons general freight, a special feature for the first time on a Union ship being a refrigerator compartment which could handle 30 tons of boxed fish… a modern ship for her day [launched in 1923], not luxurious but comfortably fitted out, with hot and cold water through forty-two cabins, two suites with private bath, and four outside rooms on the top deck with extra facilities. With a large observation room forward, an attractive main lounge and dining saloon [sic] seating sixty-eight persons…"

~

In January 2009 I drove to Horseshoe Bay and caught the 11:00 a.m. ferry to Langdale. Snow was piled high on both sides of the short cul-de-sac at the end of which lived Inga Fenwick, Good Hope's storekeeper from 1943 to 1945. As I parked, a Sheltie came down the back steps to size me up. "Rosie!" a woman commanded. The Sheltie came halfway toward me, her tail wagging, and then, at the woman's command, retreated. "Don't mind Rosie," Inga said, "she's friendly." We shook hands and, despite Rosie's reservations, Inga invited me up and into her home. We entered through her kitchen where a bright, clean, wood-fired oven warmed the room.

"I get my wood from the mill," Inga explained. "They let me take the scraps."

"Do they deliver them for you?"

"No, I load up a wheelbarrow."

I was impressed. "Do you heat with wood, too?"

"Yes, it's all wood."

She led me into the front room. On one side was a dining table and chairs. On the other side a sofa, easy chairs and a coffee table. A heater protruded from a wall, a container of wood beside it. I took a chair by the coffee table and surveyed my surroundings while Inga went back to the kichen and poured boiling water into a teapot. A shelf held a collection of books, mainly on BC ships and shipping lines. There were some paintings on the walls and a few mementoes here and there. It was a functional, comfortable, uncluttered room, but its primary feature was the clear view it afforded of the waters of Howe Sound and the forest on Gambier Island. Rosie came over and sniffed around. I patted her gently and said her name. I sensed that I was gaining a guarded measure of acceptance. Inga emerged from the kitchen with the teapot, a cup, milk and sugar.

"I won't have any tea," she said. "Just some water."

As the tea steeped, we talked. Inga was born in 1918, the first child of Mary Thompson (née Veitch) and North Vancouver lawyer Eric Thompson. She was the first girl in North Vancouver High School to take a "manual training" class. She much preferred woodworking to cooking and sewing. But in general, she hated school; it was a prison to her. She chuckled at the thought of it. "Eventually I was able to forgive my mother for sending me to kindergarten." When she was eight years old her father gave her a kindling axe for Christmas, which she thought was a wonderful gift.

> In North Vancouver when I was young there weren't very many cars around. Ditch digging was done by hand; there were no backhoes. I used to watch the men digging these ditches to lay pipe. One day I asked the foreman if I could help fill them in once the pipes were laid. He let me try my hand and I guess he thought I was okay because he allowed me to do the work. When I was growing up Vancouver was still very class conscious. Many people would put on English airs to try and inflate themselves socially. British parents would insist on their children retaining their English accents. I remember a scuffle between a bunch of us kids including the Farrells. Mrs. Farrell complained about it to my mother, pointing out that the difference between her kids and everyone else's was that "those kids are common Canadian kids, but mine are little English ladies and gentlemen."

When Inga was young, a friend of her mother's asked her what she'd like to do when she grew up. "I want to do nothing just like my mom," she replied.

As a girl she liked to watch ships load and unload at the North Vancouver dockyards, but

THE STOREKEEPERS

her deckhand days began on the dock at Hopkin's Landing on the Sunshine Coast, where her family had a summer retreat. Typically when a boat came in in those days people would gather on the dock to greet it. But for some reason on this occasion when the SS *Capilano* came in there were only a few middle-aged women and 16-year-old Inga on the dock. One of the deckhands on the ship yelled down to her, "Hey! Missie! How about tying us up?" and threw her a rope. She complied. From then on she was there to tie up the ships and eventually the captains got to know her and would invite her on board to help out. She did whatever ship work there was to do, without pay. The ships would go out in the mornings and come back to Hopkin's later in the day. She enjoyed the ships and the work. Eventually she learned the coast, and the captains would trust her to

Inga Fenwick (née Thompson) worked as a storekeeper at Good Hope Cannery from 1943–45. In the photo above she is aboard the M.V. *Lady Rose*, c. 1940. Photo courtesy of Inga Fenwick.

THE GOOD HOPE CANNERY

A "scandalous" romance. Ed Fenwick was twenty years Inga Thompson's senior when they met at Good Hope in the early 1940s. Photo courtesy of Inga Fenwick.

steer the ships. She was a paid deckhand (the only time she was paid for this kind of work) on the Imperial Oil boat *Marvolite* in 1942. The boat went from Vancouver to Wales Island and back, making sixty stops along the route over a period of two and a half weeks. On that trip the boat stopped in at Good Hope.

"The Lauritsens were neighbours of ours," Inga said. "They lived on the north side of Moody Avenue. I knew the family quite well. Elvira Lauritsen was a friend of mine." Mr. Lauritsen was standing on the dock as she finished tying up the boat and he asked her what she thought of the place. "Oh, I wouldn't mind being here," she replied. She was installed as the storekeeper the next season and met Ed Fenwick. Ed was born in Norway the year Good Hope was built and immigrated to Canada around 1923. Fenwick is an immigration department corruption of Findwick. He was a carpenter and the main "fixer upper" at the cannery, the boss of the repairs crew. During the fishing season he also worked as a tally man. Their first date was an excursion in a cannery rowboat.

During our lunch of soufflé, salad and homemade biscuits, Inga and I continued to talk about Good Hope.

THE STOREKEEPERS

The ABC company was a good company so far as I could tell. During the off-season they'd let guys live at the cannery rent-free. Otherwise these guys had nowhere to go. Housing in the war years was hard to come by.

In summer the store was a very busy place come Friday night through Sunday, a real zoo. Over two hundred fishermen would arrive wanting supplies. One time the Chinese cook came into the store, very excited. "Bossy man tell me I got $125 but I not get, get less."

"No, you also get deductions," I explained to him.

"How much I get in pocket?"

"You get $98 in pocket."

"In pocket?"

"Yes." He relaxed after I assured him that he hadn't been cheated.

Good Hope was a beautiful place, but, you know, it was marred at times by the lack of sanitation. In fact, it was terrible: sewage went straight into the bay. When I was first up there the outhouses by the Indian shacks emptied out onto the rocks at low tide. The stench was awful. Then they moved them farther out over the water. You could see everything going into the ocean. The stretch between Wadham's and Good Hope, a distance of 4 miles (6.43 kilometres), was known by the fishermen as the "shit house drift." That's how bad it was. The only place you could swim was at the mouth of the river, but it was very cold.

Good Hope's first female fish pitcher, Inga Thompson, hard at it in the early 1940s. Photo courtesy of Inga Fenwick.

The guys who worked at Good Hope in the 1940s were all older guys in their late forties and fifties, because the younger guys were off fighting the war or working in the shipyards.

Mr. Lauritsen always dressed in a suit with a bow tie and a black Homburg. I remember one time when a fisherman came in and wanted to buy a pair of pants. He didn't know his size so he tried some on behind the counter. Well, Mr. Lauritsen found out about it and told me in no uncertain terms that customers were not allowed behind the counter under any circumstances.

Another time we argued about me paying cash for goods I needed from the store. He insisted that I should run an account and have my purchases taken off my paycheque like everyone else. I told him that I'd sooner take my business to Dawson's. He wouldn't talk to me for a week. He liked to play bridge, very serious and quiet. Good Hope was a rundown sort of place. Anything that was half decent we'd say that it must have come from McTavish. Mr. Lauritsen didn't like to hear that.

Harold Wood was an Englishman. He was supposedly the cannery bookkeeper, but in fact his wife did the work. Harold told the most unbelievable tall tales. I remember one about the time he was firefighting and a bear chased him up a tree. He claimed that he fell and the bear ripped out his intestines, but he was able to tie a strip of his shirt around himself and get to a hospital, no problem. Another one was about how he stuck his head into an open elevator shaft and the elevator fell, clipping off a piece of his head; of course he'd suffered no permanent damage.

John Larsen would arrive clean at Good Hope in the spring and then he wouldn't take a bath or change his clothes for the entire season. He was hospitalized at Darby's [Rivers Inlet Hospital] one time and the nurses wanted to give him a bath, but he refused, he didn't see it as a necessity. He was a kindly man. During the war it was hard to get good chocolate, so he'd ask me to set aside the good stuff for his grandchildren.

When asked what she remembered about Ludvig Gulbranson, Inga blurted: "Haircuts! You took your chances getting one at Good Hope. One time Gulbranson cut the Chinese cook's hair. Gave him a Mohawk. When the cook saw it he quit. I remember Albert Gjertson cutting my hair. I got a 'basin cut.'"

She then went on to describe the Abelsons: "Irene Abelson was a strong, red-headed Irish woman. Her husband Olav was a very nice gentleman, quiet. I remember one time when Marea tried to interrupt her mother and another woman who were having a conversation. Her mother told her that it wasn't polite and told her to wait. Marea stood there bubbling up like the lid on a boiling pot of water."

Inga brought out a photo album and flipped the pages to her Good Hope pictures. There was one showing her pitching fish from the hold of a boat.

"My brothers Will and Jim also worked at Good Hope, as fish pitchers. My younger brother, Will, was supposed to be pitching one day, but he had broken his leg on a trip up to Sandell Lake, so I asked Mr. Lauritsen if I could replace him. He said it was okay and so I became the first female fish pitcher at Good Hope."

"And your brother, did he live to pitch more fish?"

"Oh yes, he was fine. Ed, myself and another Good Hope worker hiked up to the lake. Will was in his sleeping bag at the campsite. So long as he didn't move he was fairly comfortable, but of course we had to get him to the cannery. Ed and the big fellow took turns carrying him on their backs the four miles to Good Hope." She thought for a moment and said, "It was a beautiful lake, the moss along the beach at the upper end was thick with wild, white violets with purple stripes."

As the afternoon wound down, the conversation took a philosophical turn. "I've never had a driver's licence or owned a car, you know. We never needed a car. There were no roads where we lived. Until not that long ago there was no road to here." She paused. "You know, I never asked for anything," she said. "Things have just been offered to me."

I thanked Inga and said goodbye to Rosie. I raced and caught the 4:30 ferry. Fifteen minutes out of Langdale a fog bank loomed between us and Horseshoe Bay. Once in the fog I could just make out the peaks of the North Shore mountains, smoky blue and streaked with snow.

A big man and a big fish (tyee salmon). Good Hope manager Oly Anderson and his wife, Ilmi, c.1950s. Photo courtesy of Gerry and Rosemary Miller.

Chapter 8

The Anderson Years

Good Hope was a fiefdom, a little kingdom unto itself.

—Richard Straw, ABC Packing Company bookkeeper, 1960s

No family has more affiliation with Good Hope than the Andersons of Sointula. Their story begins with Antti (pronounced "Unti") Karvonen. Immigration officials recorded his son John as "Antti's son" and it isn't far from that to John Anderson. Around 1907 Antti's son John was working at an anvil in his workshop in Alert Bay reshaping a whaling harpoon. Unfortunately, there was still a live charge inside. When it went into the fire it exploded and John became the first patient at the new Rivers Inlet Hospital.

John went on to become a master skiff builder. He started designing and building them in 1917 and one year in the 1920s built seventy-five skiffs for Good Hope. (One of his skiffs is on permanent display at the Gulf of Georgia Cannery museum in Steveston.) John's son Oly spent nineteen years as a fish packer before taking over the reins at Good Hope from Levi Lauritsen, managing the operation from the early 1950s to its final season in 1968. Oly's uncle Ted Anderson worked at Good Hope, and his father-in-law, Gus Kallio, was a night watchman at the cannery. Oly's sons Donald and Andy worked at the cannery, too. Born in 1938, Andy's first job at Good Hope was in the store selling pop to the Native kids, the storekeeper being too busy to be interrupted all the time. Most of the Natives in the 1950s came from Kingcome Inlet. His dad also entrusted him with burning used coupons (cannery money) at the end of the season. "Whatever you do, be sure that they all burn!"

As a kid Andy rummaged around in the dirt where the China House had once stood, finding two or three hundred chopsticks, which he kept—his prized chopstick collection. He went to work on fish packers in 1954. "I could recognize a fishing boat from miles away on the open water, using binoculars from really far away. I knew them by their shapes: different cabins, sterns, varnishes on bow, paint colours. You could tell if a gillnetter had already unloaded his fish if his company flag was down. The first packer I worked on was the *Caroldale*. It could hold around 27,000 to 28,000 pounds (12,247 to 12,700 kilograms) of fish. You'd have to unload around fifteen gillnetters to get that weight. Nineteen sixty-eight was a bumper sockeye season, '73 was another big year. By the 1980s my packer could handle 45,000 pounds (20,412 kilograms) of fish. The unloading method from the 1950s to the 1960s was by pew, each fish thrown individually. It took a couple of hours to unload 5000 pounds (2,268 kilograms), 6 pounds (2.72 kilograms) per fish on average. I'd get up at 5:00 a.m. and run around and pick up fish, usually unloading by 2:00 or 3:00 in the afternoon. But some days I'd work until 5:00 or 6:00, or later. I was paid by the day, so if a day stretched way out, so be it." Andy retired in 2002 after forty-nine years in the fishing industry.

~

Gerry Miller was destined for the fishing industry. He was just three years old when a Steveston neighbour caught him as he was heading out on a big floating tree round to do some fishing on the Fraser River. Gerry's schoolmates were Japanese and he spoke their language before he spoke English. Because he seldom spoke, his mother and grandmother thought he might be a little slow. "Oh, he speaks great in our house," said a Japanese neighbour. "Well," said Gerry's mother, "I'm concerned that he doesn't eat enough." "Oh," said the neighbour, "he eats lots in our house." As far as the Japanese were concerned, Gerry was one of their own, who just happened to be white.

Gerry was working on Easthope engines before he was twelve years old. He gillnetted from 1946–50 when George Skinner, the skipper of the ABC packer boat the *Fir Leaf,* asked if he'd like to be the boat's engineer. That led Gerry to Rivers Inlet in the winter of 1951–52. Gerry's friend Cecil Fisher was the shore engineer at Good Hope, but he didn't want to do it anymore, so Gerry took his place. Good Hope had bought thirty old gillnetters for a few hundred dollars each from the ABC cannery in Skeena River. They were in bad running shape, and no one could—or wanted to—fix them. Gerry arrived and got them all going again. One had to be resourceful in those days, so Gerry would take a transmission from a 1934 Chevrolet, marry it to a differential from a Model T Ford, and fashion a 1⅛ shaft suitable for the drums on the boats. He did this kind of repair work right up to the early 1960s.

> There was a Finnish fisherman, Vino Helminen, a strange character, never washed; his clothes were dirty and shiny from use. They used to drag him into the sauna and get him cleaned up. He didn't like to put money in the bank because he knew the government would get some of it, so he never deposited his cheques, which used to drive the company crazy because they could never clear their accounts. Vino had an engine on his boat that wouldn't run properly. He went to Oly Anderson and said, "I need a new engine."
>
> "What's wrong with the old one?"
>
> "It's not going to bother me anymore." He'd thrown it overboard.
>
> So they put in a new four-cylinder engine with an exhaust pipe, but because it was so long it got condensation

Silver on its way to becoming gold. Cecil Fisher on the ABC Company packer *Pine Leaf*, 1940s. Photo courtesy of Marlene Yurichuk.

THE GOOD HOPE CANNERY

Rivers Inlet nets were a frosty, light green to match the colour of the water and be invisible to salmon. Dick Haggerty on his boat *Roseanne H.*, 1950s. Photo courtesy of Gerry and Rosemary Miller.

in it. Vino started monkeying around with the engine, drilling holes in it and everything, and finally I got tired of him coming around to me. He had wires all over the place and so I built a dashboard and mounted the gauges there with various toggle switches, and one switch that didn't connect to anything that I put a bunch of wires around to make it look like something. I explained to him what everything did. "If you're having trouble with the rotinators just flip this switch and you'll have no trouble." That was the unconnected switch. Well, he came in at the end of the week and he said, "By God, did that switch ever work!"

I was working on a boat one Sunday afternoon. Everybody was starting to go out fishing and this Japanese guy came along and said his engine wouldn't run properly. And so, even though I was busy at the time, I said okay, I'll come now. Well, he started mumbling in Japanese, "Damn white guys… hit them over the head and throw them in the chuck." This guy had a small axe just inside the cabin door, so I was a little worried. He was saying "I'll throw you overboard" and stuff like that. Of course, I knew what he was saying, but he didn't know that I knew. Anyway, he wouldn't stop so I threw him over into the salt chuck. A bunch of people were watching and they said, "That was beautiful, Gerry, that was beautiful!"

Dick Haggerty, a fisherman and watchman at Good Hope, married Annie, a British war widow with a young daughter named Rosemary. Dick used the money he got from the

THE ANDERSON YEARS

government for his war service and bought a fishing boat, but it wrecked on some rocks and he took a job at Good Hope as caretaker. "Our house in Richmond was being raised to put in another floor when we got the phone call in the late fall of 1961 that Dick had drowned," recalled Rosemary. "He'd been over to see Lucky Bachen at Dawson's, to get the mail. He'd tied up his boat back at Good Hope and fell in the water by accident. He had rubber gear on and he had emphysema. He tried to swim but he was too tired and he drowned. They eventually found him under the cannery wharf, his knees all red from the copper paint on the bottom of the boat."

Rosemary ran the store for a number of seasons and met Gerry. "I was miserable my first year up there. I was eighteen years old, there was no radio, the lights went off at eleven o'clock." Gerry made it more tolerable for her and they eventually married. Rosemary was pregnant with their first child when she went into labour at Good Hope. A float plane arrived to take her to Bella Bella, but as they were taking off a log came shooting out of the Sandell River. The pilot had just enough time and speed to hop the plane over the log, narrowly missing it. Nervously, the pilot smoked a pack of cigarettes as they travelled the forty minutes to Bella Bella. When they arrived he dropped Rosemary at the wharf, turned around, and took off again, leaving her standing there. A week after having the baby, she bundled the boy, suffering from jaundice, in a green blanket and set off for Good Hope in

Its canning days over, Good Hope continued to serve its fishermen as a place to clean, store and repair nets. Photo courtesy of Gerry & Rosemary Miller. 1950s.

a fish packer. When she arrived back, Gerry took one look at the tiny, slightly yellow baby in a green blanket, and said, "Is that it?" "It" was Kenny, but despite this less than graceful entrance into fatherhood Gerry and Rosemary had two more children. Gerry's working hours were very long and at times Kenny would ask, "Have you got the day off tonight, Daddy?"

"I remember some Indians finding a baby eagle at Good Hope," said Rosemary. "I fed it stew meat. After the Indians left, Pete Lindstrom looked after it through to maturity. He said that it came back once or twice after it was mature and circled back and he never saw it again. It stayed around for about a year. It started to forage farther and farther as it matured. It came back after a year and sat on a pole outside Pete's house.

"The young fish pitchers were always complaining that there was nothing to do and Gerry couldn't understand it. I looked out one day and there was a gillnetter going around the bay with a man behind it being towed on a door. It was Gerry."

"You didn't have to be going very fast," added Gerry. "You'd create waves going around and around and hop over them. Andy Anderson drove the boat; all the kids got into it."

"The difference between the Norwegians and the Finns," Rosemary explained when asked about the subject, "was that if a group of Norwegians were standing around talking and you came by they'd switch to English, but the Finns wouldn't."

When asked about her memories of the Natives, she said, "At Alert Bay the Indian women would ask, 'Is it the weekend today?' Because the weekend started whenever the fishermen finished fishing for the week. It might be a Tuesday or a Wednesday, and so that was the start of the weekend for them."

Gerry was reminded of Mr. Chiba, a boat builder at Good Hope.

Mr. and Mrs. Chiba lived in the house at the end of the bay at Good Hope, the former Abelson house. Chiba was a boat builder. He worked from raw cedar, planing everything down by hand. It took him two years to build his boat at Good Hope.

Fred Kwan was one of Good Hope's cooks in the late 1950s, a good baker. Mr. Wong was another cook in that era, his last year was in '53 or '54. He constantly fought with his kitchen helpers, what he called "flunkies."

THE ANDERSON YEARS

The bookkeepers at Good Hope from the 1950s were Ernie Hill, Phil Beasley, Ivan Ketchen and Wally Lehti. One day Wally decided that he was going to get rid of all these old accounting records that they stored in the attic above the office. So he took wheelbarrow after wheelbarrow full of papers and dumped them in the fire pit of the boiler. Well, the boiler got so hot that the water main started backing up. Nancy Anderson [manager Oly's young daughter] just happened to be walking down the pathway from the bunkhouse to the office when a bung blew out of the water line, startling her and stopping her from going forward.

Mr. Chiba's boat, hand built from bow to stern over the course of two years, was engineered by Gerry Miller. Note the ABC flags, c. 1952. Photo courtesy of Gerry and Rosemary Miller.

151

THE GOOD HOPE CANNERY

She turned around to go back and another bung behind her blew out, so she was trapped with hot water spraying out everywhere. Her mother Ilmi heard her yelling and came to the rescue. Anyway, Nancy was okay, no harm done.

Hayden Jones was a storeman in 1951. He was later well known as singer "Johnny Hayden." The net bosses were Gjertson, Ed Thuvesesn, Otto Wright and Phil Hilton, and there were a couple of old fishermen who'd always show up at Good Hope to fish from skiffs. Anderson would be after them to not come up anymore.

Gerry's last season at Good Hope was in 1963.

~

There was another John Anderson who fished Rivers Inlet for many years, including one season in the early 1950s with his grandfather, Hans Peterson. He had three uncles who fished the inlet for ABC: Peter, Ivan and Elmer Peterson. John fished one season for ABC and spent twelve years with Wadham's because manager Hans Otteson set him up with a gas boat that John named the *Sanpo*. He was known for fishing Wadham's Point, but he recounted having fallen asleep one night early in his career while fishing at the head of the inlet. He was alone and woke up to discover that the net was twisted fifteen or twenty times around his skiff. He managed to haul this clumped mess of linen net and lead lines over the roller and back into the boat. At around midnight he started rowing.

Running the kitchen at Good Hope was a demanding job. Cooks came and went with regularity. Pictured: Mr. Wong. c. 1953. Photo courtesy of Rolf Hundvik.

After a trip of about ten miles, he arrived at Good Hope around 6:00 a.m. and "went to sleep in one of those shacks they had, just a straw mattress and a lantern for a light. I got up a few hours later and went back out."

John recalled that "In the late 1960s you could make big money, $10,000 a season," but he called it quits with fishing in the 1970s. "Vietnamese fishermen started appearing in the inlet then. They were not pleasant to deal with. The federal government sponsored and subsidized them. If you argued with them they would bring out guns, AK47s." Good Hope fishing guide Bill Hobbs recounts a similar incident from the same time period: "There was an old gillnetter by the name of Ray Reese. One day a seine boat drifted into his net. Ray warned him off, but the captain of the seiner, a Yugoslavian, brought out a gun in response. Ray, who had been a paratrooper in the Second World War, retrieved his rifle and brandishing it, challenged the seiner captain, 'Shooting? Killing? That's my business!' The seiner captain backed his boat away."

I spoke with Rivers Inlet/Good Hope fisherman Sam Macki in March 2009. Retired since 1998, the jovial 78-year-old was three times a "high liner" on the coast. I asked him if he considered himself an expert fisherman. He laughed pretty hard at the question. "I remember Oly Anderson saying to me one time, 'Macki, just remember, you're not an expert, you're a fisherman.'"

> I was born in Vancouver, east end, and learned to speak Italian before I spoke English. Both my parents were Finns. They came to America—Canada or the US was America to them—because they believed the streets were paved with gold. I started school in Ladysmith and then attended Seymour-Strathcona school in Vancouver. Eventually we settled in Sointula. [His first year fishing was as a 13-year-old in 1944.] We got off the boat and onto the dock at Good Hope... we were about to go up the gangplank when we spotted a couple of cute young girls [Mildred Lauritsen and Alma Gulbranson] standing at the top... well, we jumped in a dinghy and rowed to the other float rather than have to face those girls... we were shy young guys! I worked on the *Island Maid*... the vibration from the old two-banger would tickle the hairs in your nose; she was the only boat I was on that ever did that. The first meal I ever ate on board

a fishing boat was Lipton chicken noodle soup mixed with Lipton tea. I ate that "meal" with Oly Anderson. We were terrible cooks… I remember trying to make rice… we put in so much rice that as it cooked the pot kept overflowing with it… I fished with a guy we called "Stinky" Mason, he smelled so bad they wouldn't let him on the boat; they towed him behind in a skiff. There was a bootleg joint on a float across from Dawson's where the men would drink. We were just kids and got tired of waiting, so we took a boat out for a joyride. The fishermen weren't too happy with us. Fishing to me was like hunting. It wasn't about the dollars, it was like stalking a deer. I'd get so keyed up the first day out that I couldn't sleep that night; I was so excited. Catching fish is a matter of common sense and intuition; it's a feeling. If I couldn't afford to take chances one year I'd go where I knew I'd catch fish, but if money wasn't a concern I'd play my hunches. I learned to swim in English Bay and Second Beach. When I wasn't fishing I used to go swimming, diving off the dolphins [three pilings tied together] in Goose Bay, or off the bridge of the boat. *Velmar* was my boat.

Not every fisherman knew how to swim, sometimes with terrible outcomes. "Paddy Henry, an Indian guy, drowned not far from Good Hope in the mid-1950s," Sam recalled. "The boat was running and he fell out. He yelled at his five-year-old son to turn the engine off, but the kid didn't know how, and Paddy drowned." That sad tale prompted me to ask him about any fisherman superstitions. "Oh yeah, there were a few, such as you had to open a tin of Pacific canned milk with the label facing up, otherwise you had to throw it overboard; you weren't supposed to bring flowers on board; whistling was bad luck because it meant you'd bring on a storm; some fishermen thought it was bad luck to leave port on a Friday, so they'd wait until after midnight if they had to. And of course you never brought a suitcase on board, you packed your things in garbage bags—fisherman's Samsonite we called it. I'd still be fishing if DFO hadn't screwed up the industry beginning in the late 1980s. We all used to be on a first-name basis, but then it all changed. DFO officers started carrying arms; bulletproof glass in their office. You couldn't talk to them anymore. One time they came on board my boat without my permission and went through my logbook. That's not right."

~

Glen Evans, son of long-time H. Bell-Irving & Company bookkeeper Cyril Evans, went to work at Good Hope as a 15-year-old in 1962. "It was a net loft when I worked there. I was the gas attendant. I had to row across the bay to the gas barge, a distance of about two hundred yards (183 metres). Work consisted of filling up fishing boats, checking supplies, hoisting and manoeuvring big drums. I'd row back for lunch, and then out again, and back after work. I slept in the bunkhouse."

In subsequent seasons Glen worked on the collector boat, the *Miss Nancy*.

We went out twice a day. The skipper was Andy Anderson, Oly's son. The fishermen didn't own their boats, they were financed by the company, so Oly had to keep a close watch on them and their whereabouts, make sure they were fishing when they were supposed to be. Most of the Indian fishermen seemed to be from Alert Bay and some of them had a tendency to go down there, about a day's run from Good Hope, to drink instead of fishing. There were not many places where you could get alcohol. The only other place where it was available was in Ocean Falls. Oly's biggest task was keeping control of the Indians, making sure they were catching fish. During the fishing season the Indians lived at the far end of the property. Oly warned me not to go down to that end of the camp. I was told to just leave them alone, that if I left them alone they'd leave me alone, that was all. But I was friendly with the Indian kids my own age; we used to go fishing together up at Sandell Lake.

The Japanese owned their boats, so there was no leverage over them. They were all business, and the company had to keep them happy to keep them fishing for Good Hope. The Finnish fishermen had to have their sauna, so the cannery had a floating version—a 45-gallon [200-litre] oil drum that had been cut in half and filled with hot water—that they'd tow out to the fishermen. It held six or seven men at a time.

There's one fisherman that I'll never forget. His name was Paddy Flynn, a big, temperamental Irishman. I happened to be working in the store one day when Flynn came in and ordered groceries, including steak. I was about to wrap up a big steak when Flynn ordered me to give it to him, "Right now, just as it is." He proceeded to eat the steak raw while I put together the rest of his groceries. I almost threw up.

THE GOOD HOPE CANNERY

I was out on a boat with Oly one day when he spotted a deer on the shore. He took his rifle and shot it. I had to row ashore to retrieve it. It was still alive when I reached it, so I had to slit its throat. I had no idea what I was doing, but I did it. Somehow I got the deer into the dinghy and rowed back to the boat. Oly complimented me on the fine job I'd done.

Fishermen of Finnish descent were integral to Good Hope's success. John Grohn began fishing Rivers Inlet for ABC in the '30s. Here his son Marty pitches a fish into the hold of a packer boat, c. 1968. Photo courtesy of Marty Grohn.

In March 2010 I met Marty Grohn, who fished three seasons—1967 to 1969—with his father, John, in Rivers Inlet, catching fish for ABC with the *Kathy Lynn*, a boat he bought from Dick Haggerty. "My grandparents were from Helsinki, Finland. They were tailors, made uniforms for the Vancouver police and fire departments, but my dad wasn't interested in tailoring. He took up fishing instead. He started in the 1930s in Rivers Inlet. They'd raft together going up there, tie three or four boats together in ideal weather, one guy would steer and the others would drink whiskey. The Finns had a camp near Duncanby Landing in Hakes Bay. They had a sauna bath at the end of the dock there."

John Grohn's *Pearly Shell* seine boat was chartered out to ABC in 1947. He built the *Sedrik* in the back of a business at Hastings and Penticton streets in Vancouver and launched it in 1953, then went on to build the 47-foot (14-metre) seine boat *Estric* in 1953–54.

In 1962 Richard Bell-Irving's son Ian took over as president of H. Bell-Irving & Company. Ian had joined the company in 1946 and worked summers in ABC plants before becoming manager at Arrandale Cannery on the Nass River. He inherited a business under tremendous stress in a time of sweeping cultural, political and economic change. By the 1960s it was clear to the federal government that there were too many boats and not enough fish. In 1968 Jack Davis, the federal fisheries minister in the newly elected Trudeau Liberal government, announced that the open-access salmon fishery would become a limited-entry fishery, effective with the 1969 salmon season. Known as the "Davis Plan," the new policy set out a two-level system of licensing based on salmon landings in the two previous years, and included later phases that provided for fleet reduction through a buyback program and licence retirement. What had been five days of fishing became two or three and fishing licences that had been worth $100 went up to $10,000. Fines for dumping fish offal into the ocean came into force, leading to increased costs for canneries. Native fishermen began to be in charge of their own destiny. Twenty-four of twenty-five seine boats fishing for ABC's Alert Bay Cannery were Indian owned and operated. "Under [Ian's] presidency," writes K. Mack Campbell in Cannery Village: Company Town, "the company sold the Fidalgo operation and in 1966, tried diversification into the herring reduction business in New Brunswick. With bank

THE GOOD HOPE CANNERY

financing becoming tight, Anglo-B.C. sold its northern B.C. operations to Canadian Fishing Co. In 1968, the Phoenix plant on the Fraser River was sold to Canfisco and the Anglo-B.C. seine fleet was sold to Nelson Brothers Fisheries."

Richard Straw witnessed the end of ABC. Born in Vernon, BC, the 21-year-old went to work for the Bank of Commerce and in 1966 was sent to their Alert Bay branch. There he met Bob Sherwood, the bookkeeper for ABC's Alert Bay Cannery. Bob offered him a job as a bookkeeper and he accepted.

> Norm Corker was in charge for ABC; everyone knew Norm. He was known up and down the coast simply as "the boss." I was looking for adventure, and working for a cannery fit the bill. It was definitely a work-hard, play-hard situation. During the fishing season we would work 24-hour days during the week. Weekends we partied, camped, fished and hunted, or otherwise explored the coast. The coast was the place for oddballs, but they were tolerated. I'd work at ABC's Phoenix plant in Steveston in the winter, and up the coast going from one ABC cannery to another in the summer. John Cormack was an ABC fisherman and captain of the *Phyllis Cormack*. It became the first Greenpeace protest ship, the *Rainbow Warrior*, and he was its captain. In about 1966 the Vancouver rock 'n' roll radio station CKLG increased the strength of its signal—up until then you could only get CBC—and you could pick up CKLG if you went out into the Johnstone Strait away from the mountains. Good Hope was a fiefdom, a little kingdom unto itself. Oly Anderson ran it and Marvin Jones was the bookkeeper. He was a religious man, involved in the Salvation Army; he played in the band.

In the fishing business the band was playing a new and different tune, but the ABC head office in Vancouver was tone deaf. George Griffith was the Bell-Irving/ABC head accountant. "Griffith was a textbook old-school guy, very patrician," said Straw. "We'd show up in our T-shirts and sneakers from being up the coast and they wouldn't like it at all. There was a young receptionist in the office named Hadee. One day she showed up in a miniskirt and nearly gave the guys a heart attack." Peter Traill, who had become secretary of the company in 1916 and was Richard Bell-Irving's mainstay was, as K. Mack Campbell writes,

THE ANDERSON YEARS

"the benevolent 'eminence grise' of the company, and [was] largely responsible for its ultra conservative approach."

Ultra-conservative or not, by the 1960s the business was going broke. The Fidalgo Island Packing Company operation was sold in 1964. ABC continued to run four big operations in BC—Good Hope, Alert Bay, North Pacific and Phoenix. In 1966 the company thought it saw a profitable future in the East Coast herring reduction business and bought a plant in Caraquet, New Brunswick. But in BC things continued to slide. "The big mentality within ABC was to catch as many fish as possible and pack them," said Straw. "You dealt with the money later. Everything went into boxes to be sorted out later. Accounting was done by pencil. I remember buying a hand-cranked adding machine; that was a big innovation. Money was spent without regard. There were virtually no controls on spending. The company would finance up to 200 percent of a private fisherman's boat. The accounting was terrible. It seemed to me that there were two sets of books, one for the fishermen and one for the company. They'd pay certain fishermen wintertime salaries for basically doing nothing, to ensure that they'd be back fishing for the company in the summer. They didn't charge interest on accounts receivable. It was the old indentured system. Poor old ABC was completely out of touch. There was lots of sentimentality for the old fishermen and the old way of doing things. Costs were hidden from head office. In contrast, BC Packers was on top of things daily. Finally, in about 1968, ABC brought in an Australian chartered accountant, Mel Newth, to look at the business. He spent two weeks visiting the canneries, looking at everything. At the end of it he told the company 'You're spending money faster than you're making it, you guys should be in receivership. It's too late to turn things around, you should shut down the company.' Everything had been done on trust up to that point, so anyone owed money by ABC expected to be paid as always, but the reality was that the company collected whatever money it could and then shut down. A number of people were owed money at the end. There were some bad feelings about it."

Ian Bell-Irving's vision was off to a good start. Guides Bill Hobbs (middle) and Jim Frost (right) and a big catch of coho, 1970. Photo courtesy of Bill Hobbs.

Chapter 9

Hopeful

[Good Hope] stood on stork-legged piers amidst steep mountains halfway up Rivers Inlet, one of those fog-shrouded fjords of the British Columbia coast that make of the province a kind of super Norway... Its clapboard siding was weathered bone-white by the North Pacific winds, its seams shrunk tight with cold and brine. The pier pilings, crusty with barnacles and mussels, supported Augean sheds and mammoth barrels stained with the blood of countless dead salmon; the blank eyed buildings stared toward the west with a gaze as hard and cold as the weather itself. Over the whole scene hung the iodine odor of kelp.
—Robert F. Jones, *Sports Illustrated.*

The Anglo-British Columbia Packing Company sold its salmon cannery business in 1969, ending the company's operations in BC. With the closure of its herring reduction plant in Caraquet, New Brunswick, in 1974, ABC ceased to exist at all. In the midst of this, Ian Bell-Irving hung onto Good Hope.

"After we sold the company nobody wanted Good Hope, yet it was in mint condition," said Bell-Irving. "I'd gone up to Glendale Cannery in Knight Inlet to see this chap from Campbell River who was interested in sport fishing up in that area, and I suddenly got the idea that maybe we should convert Good Hope into a fishing lodge." His original intention was to use it as a private retreat for friends and business associates. At the same time Rivers Inlet was becoming a playground for wealthy Americans such as Walt Disney and John Wayne, who came up in their yachts. For years Boeing executives and their wives came to fish, but the women didn't like being stuck on board for days on end. Boeing approached Bell-Irving and

THE GOOD HOPE CANNERY

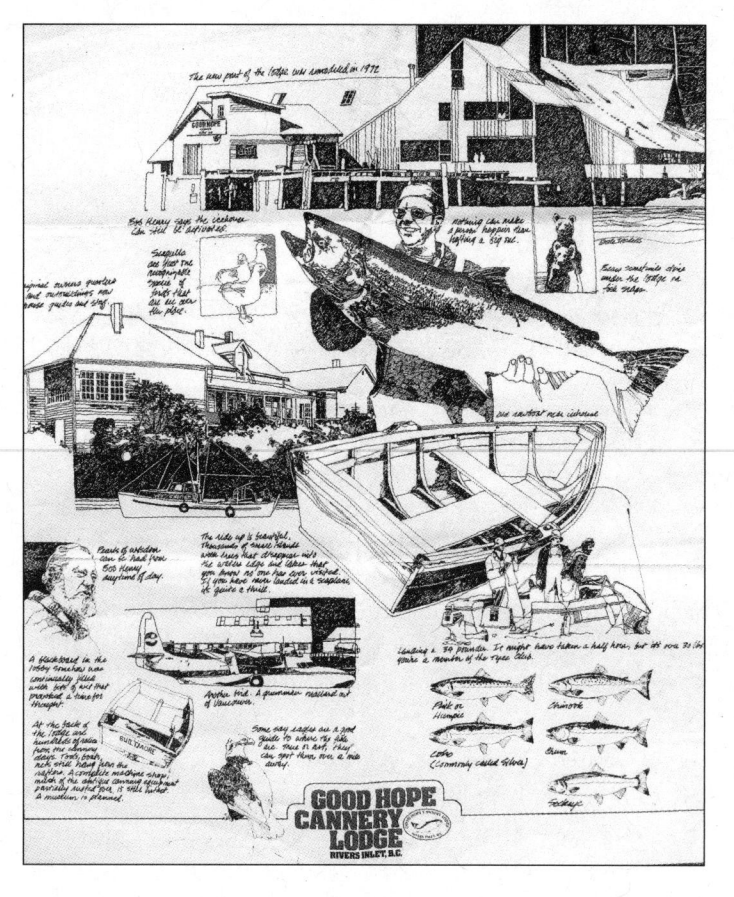

This advertisement for the Good Hope Cannery Lodge shows pilot Bob Henry (illustrated bottom left) and restaurateur Jeff Crawford, who bought out Ian Bell-Irving in 1973. Courtesy of the Richards family.

asked if Good Hope would accommodate them. Bell-Irving saw it as a business opportunity and opened the lodge to them. His thinking evolved. The lodge would cater not only to fishermen, but to people wanting to go on ecological or photographic tours. At the same time Bell-Irving bought a resort property on Saint Lucia in the Caribbean. The idea was to operate Good Hope in the spring and summer and Saint Lucia in the fall and winter.

"Ian and Joan Bell-Irving thought and acted younger than their ages," says fishing guide Bill Hobbs. "They were sincerely interested in people. Ian loved Good Hope and would join guests and staff in the bar. Joan was lively and energetic; everyone liked her."

The Bell-Irvings hired Vancouver architect Bob Hassell of Hassell Griblin to design

HOPEFUL

and oversee Good Hope's conversion to a lodge. When the Bell-Irvings asked him to take on the project, he thought immediately of his maternal grandparents, the Vicks, who settled in 1910 at Cape Scott, on the northern tip of Vancouver Island. Bob's grandfather James Vick would row to Rivers Inlet for the start of the fishing season, the only cash work he would have for the year.

"I don't remember what the budget was," said Hassell, "but it was next to nothing. We had to keep it simple. The cannery had a barn-like charm and historic value, and we wanted to retain that big roof. There was never any question of changing the basic structure. We started work in the spring, and we were really lucky because it didn't rain. We put in the stairwell and the skylights, but otherwise kept it pretty much as it was structurally. We tried to use local materials as much as possible. We got our cedar timbers from [Fred Wastell's lumber mill at] Telegraph Cove. Roy Hayward was the foreman on the job. We hired workers locally as much as possible, including some Oowekeeno men. I'd go up every week or ten days to check on their progress."

One of the workers was 20-year-old student Michael Broughton, about to enter his third year at university. "I arrived at the terminal on the Fraser River and joined Bob Hassell, the architect, and two carpenters for the trip to Good Hope." Henry Smeets was their pilot. "Smeets was Dutch," said Hassell, "from a Dutch colony in the tropics somewhere I believe, and he had dark skin, he looked like a native, and the Indians took him for one of their own."

> I was crammed into the back between the two carpenters, a drum of something on the floor between my legs. The Cessna was overfilled with us, supplies and the carpenters' tools. We skimmed along the water but the plane couldn't get up, it kept struggling to get in the air. Finally, a boat was passing in front of us and the pilot used its wake to give us the lift we needed. Our flight to the cannery was uneventful, but once we reached our destination the pilot put the plane into a steep corkscrew and sort of bounced us into Good Hope. It was so sudden and steep that one of the carpenters threw up.
>
> The foreman [Hayward] was a Brit, a guy in his sixties. There were three or four carpenters in total. I went to work scraping, carrying lumber, painting. I had to spray dark brown paint.

THE GOOD HOPE CANNERY

I wore a mask and goggles, but the paint got around my eyes. It was too hard to take off every day so I just went around all the time like a raccoon. We worked ten-hour days for six days straight every week for three weeks. One day one of the carpenters fell off a ladder. He knocked the wind out of himself pretty bad and injured his head. It took about an hour to get a plane in there and get him to a hospital. He lived, but we never saw him again.

The caretaker had a dog and a boat that looked like a tugboat. While we were up there he ran the boat into an island hard enough that he managed to knock down a tree. The tree fell onto the boat and killed his poor dog.

Ian Bell-Irving was up there overseeing things. He was a dapper-looking guy, like he'd just stepped off a yacht. He'd had some specially built tables, steel frames and varnished wood laminate, shipped up to use in the lodge. They'd just left them on the dock where it had rained on them. Plus they'd been banged up and chipped. Of course no one would take responsibility for the negligence. Ian was pretty upset, for good reason.

I'd been up there for almost three weeks when my girlfriend, Brenda, sent me a telegram. "I love you... I miss you..." and so on. Well, at that time, unknown to us, telegrams were broadcast on "the sched" [schedule] over the radio at 8:00 or 9:00 at night so everyone on the coast heard this embarrassing lovey-dovey message.

We had just finished the work when the first group of fishermen arrived. They were Americans with all the latest gear and so on. Well, one of the Oowekeeno workers took a look at these wealthy Americans and muttered, "Is this why we built this place? For guys like this?" and put his hand on his knife. One of the other workers, a real tough guy named Harry Anderson, had to tackle him before he could get to the fishermen. There wasn't much to do at night, so the crew of workers would take on little projects around the cannery to keep busy. Roy Hayward went to work on the Iron Chink and got it working. There were rooms upstairs in the cannery that were locked. The guys would open them up to see what was in there. They found cans, labels, all sorts of stuff tucked away. We even did some canning, put labels on them. There were a lot of little things that were interesting about the cannery, like the Japanese bath, a clever design. I remember a railing inside the cannery that had been worn smooth and burnished a deep brown by countless fishermen with fish grease on their hands. Every once in a while there'd be a problem with the water line, a tree might have fallen on it or something like that. So somebody would have to follow the line up the creek and make the repairs. Well, the grizzlies used the same route, so the night before they'd break out the rifles from the gun locker and draw straws to see who'd be the "lucky" ones who had to go. We'd go and look around the Beaver Cannery on the other side of the inlet—an amazing building. It was a ruin by then, but it was still a wonderful structure, like a museum. There was a double-ended troller inside, like new, a gem. We'd go over to Lucky Bachen's. Lucky had a cat that was notorious for getting on fishing boats, so whenever you docked you'd have to post someone to watch out for the cat, otherwise you'd be making a quick return trip to drop off his cat. At Good Hope Ian Bell-Irving tore down what was left of the walkway and the shacks that ran down to the head of the bay. He didn't think they looked very good, and maybe he was right or maybe it was an insurance issue.

Our work done, the same pilot arrived to take us back to Vancouver. When we were in the air he casually told us that he wasn't sure how much gas he had in one tank, but that he liked to run it dry anyway before switching over to the other tank. It turned out that he didn't fly using his instruments because they used electricity and that required a magneto, and his wasn't working. As we were approaching Vancouver I asked him if he could "drop" me off closer to home, somewhere in Burrard Inlet would be great. "Do you see any log booms down there?" he asked. "Yes," I replied. "Okay," he said and down we went. I got out onto a boom with my suitcase in hand and he took off again. So there I was on this log boom hoping that it actually stretched all the way to dry land. Somehow I made it to shore without falling between the logs. I picked up my paycheque at the ABC company office in Vancouver. I can't remember how much it was, but I remember being surprised that they deducted my room and board.

In 1970 Bell-Irving ran ads in the Vancouver papers looking for kitchen, bar, maid, maintenance and fishing guide staff. Among those he hired were four friends: Barb Sutcliffe, Wayne Kinnear, Brian Naphtali and Bill Hobbs. Almost forty years later I ran ads in the Vancouver papers looking for anyone with a connection to Good Hope. Dozens of people responded including Barb Quinn (Sutcliffe). She contacted some of her old friends from Good Hope and graciously arranged a meeting at the home in Vancouver that she shares with her husband, Russ. Barb recalled two six-month "stints" at Good Hope.

We were attracted to the Bell-Irvings' dream of renovating an old, remote commercial cannery into a fishing resort. I'd just completed the Cordon Bleu course in France and was hired

One of Good Hope's original guides, Brian Naphtali (right), and happy guests, 1970. Photo courtesy of Barb Quinn.

as sous-chef. My boyfriend at the time, Wayne Kinnear, was hired as a fishing guide along with Brian Naphtali and Bill Hobbs. As the seaplane swooped over the small group of rustic buildings, we were energized by the sheer excitement of living at Good Hope for six months. Opening a camp offered dozens of chores. We were all employed to unpack supplies, reorganize bedrooms and laundry, sew new bedspreads and throw cushions, equip runabout boats, learn to read tide charts, check first aid supplies, et cetera. We were a team of about twenty that first year including the caretakers, Rusty and Shirley Harrison, and their young twins, Jesse and Cory. My split-shift job was to assist the cook, "Scoop," in preparing meals and making bag lunches for the fishing guests.

Weeks flew by. Ian and Joan were wonderfully hospitable hosts to both the guests and staff. Ian's wink or grin was partial to cute young gals. I had enough attention from Wayne. I could daydream for hours. I loved the simplicity of walking a short distance from our bunk room along the worn wooden boardwalks in the early morning light, smelling the cedar trees and watching the bats return from their nightly hunt. I thrived on learning to batik scenes of fish and developing pictures in the darkroom Shirley set up for all to use. It was so stimulating that time disappeared into thin air. It was a living dreamtime I will never forget. On one of our days off, Wayne and I loaded an aluminum runabout with a picnic and fishing gear to explore the beaches near the mouth of the Fitz Hugh Sound. We were in wide-open water when we were so awestruck, Wayne turned off the motor. Our jaws dropped and our hearts beat with wild joy. There ahead was a hill of ocean water bulging with life. Herring chased by northern coho flying up into the air; eagles swooping

A recently graduate from France's Cordon Bleu cooking school, Barb Sutcliffe was Ian Bell-Irving's choice for sous-chef in 1970. Photo courtesy of Barb Quinn.

THE GOOD HOPE CANNERY

down to grab the salmon with their powerful [talons]; dolphins dancing in and out of the luminescent sunlight, splashing rainbows of water; seagulls circling and diving into this rising mass of herring and salmon. It was an exquisite feeding frenzy graced with movement like a finely choreographed dance. We were part of it, right in the middle of it. Wayne caught some of it on film. I remained wide-eyed in absolute delight. I'll never forget how quickly this scene erupted, both its stunning order and wild chaos. It was equally sudden in its return to regular windswept waters. As we approached the protected waters of a wide bay, we noticed humpback or gray whales playfully weaving through the dark waves. In this remote location, nature dazzled us with an abundance of beauty, sounds, clean air, surprise, food and space.

Wayne and I were in our element at Good Hope. Being Irish, Wayne loved to spin stories and tell jokes. He had the ideal captive audience in the small fishing skiffs. He'd elaborate a client's experience of catching a large salmon with just the right amount of drama so the men would swell with pride and be thirsty to fish again the next morning. The guides received substantial tips if the angler was successful. Wayne paid attention to the weather, where the fish were active, having the right tackle and avoiding snags. Wayne's tackle box was heavy with lures and spinners, hooks and weights, lines and extra reels. He strapped a sharp fish knife on his belt and wore a heavy plaid jacket. By the end of a day, Wayne smelled like cut-plug herring and fish blood. His favourite place at night was near the bar drinking Scotch whiskey neat and smoking Camel cigarettes with his new fishing friends.

Another vivid memory I have is being on one of the teams fixing the water line. Our trusty maintenance man, Rusty Harrison, was a genius with small bits of plastic, duct tape and pieces of hose. When water stops supplying the camp, every spare hand is required to help with repair. I recall scrambling up the mossy landscape, climbing over fallen logs and lugging supplies. It was often hard to hear over the raging river so we grew accustomed to using all our senses. Often it was a messy problem needing a smaller pipe or ropes or standing in the swift river. This day I watched Rusty unclog earth by improvising a tool from a root. We knew our rapid response was necessary to prevent further damage or to ensure our drinking water didn't get contaminated. It was hard, intense work calling on super strength and mindfulness. Thank goodness I was never asked to unclog a septic tank. Those two six-month periods of my early twenties remain dreamy. While those stints at Good Hope were the best job for my adventurous soul, Wayne and I parted company after we returned to Vancouver.

Bill Hobbs had learned to fish at Qualicum and had already guided in Campbell River. He would guide at Good Hope in its first two seasons, 1970 and 1971.

The feeling of being in the lodge when the rain was just pounding down; I've never seen it rain like it does up there. We would dip into the water with our nets and fill them up with herring. They were so plentiful and it was that easy. On one trip we caught something like twelve salmon. One of them was in the 40-pound-plus [18-kilogram] range, two were in the 30-pound [13.5-kilogram] range, a couple were in the high 20-pound [9-kilogram] range, and the rest were coho. It was getting late in the day when a big storm came up. We headed for the nearest shelter; it might have been Goose Bay. We were warmly received. They called DFO and they sent their big boat over to tow us safely back to Good Hope. That's just the way it was up there: people would help each other out. Another time I was guiding a guest by the name of Ganch. We were up at the head of the inlet, fishing for coho. Mr. Ganch had a beautiful rod, very expensive, with silver eyes. He had a fish on his line but somehow lost his grip on the rod and into the water it went, straight down. A little while later another fisherman comes alongside our boat. He thanks us for his fish that we caught and hands us back Mr. Ganch's rod, perfectly intact. He'd snagged Ganch's line and had pulled the whole thing up, fish still on the line.

John Salo [a local logger from Sointula] had this way of ramming his feet hard into his boots. So day one of the waitresses, Anne, put a little surprise into John's boot. He sat down, picked up his boot, and rammed his foot into… a big piece of pie.

One year in Rivers Inlet I was working as a cook on a fisheries patrol boat. The officer and I, armed with rifles, went up a stream in Draney Inlet. Our job was to hike miles up the streams removing debris that might inhibit spawning salmon. We got deep into the forest when we noticed these rubbings on the bark way up in a tree. Then we saw salmon carcasses. Sure enough, about two hundred feet away, a grizzly stood up and sniffed the air. She had two cubs with her. It was an incredible sight to see; I'll never forget it. We slowly backed away, hoping we wouldn't have to use our rifles, and we didn't.

Hearing this, I couldn't help thinking of Henry Bell-Irving and what he would have given for a shot at this bear.

THE GOOD HOPE CANNERY

As I was leaving the Quinn household Russ mentioned—in yet another Good Hope small-world moment—their friends Charlie and Patricia Wilson. "Patricia was a Bell-Irving," he said. "She loves to talk about the Bell-Irving clan; you should give her a call." In May 2009 I was welcomed warmly into "Trish" and Charlie Wilson's home. Trish and I talked for hours while Charlie, who assured me that he'd heard it all before, watched television.

Trish's father was Allan Duncan Bell-Irving, the fifth born of Henry and Isabel's six sons. Her mother was Mary Elizabeth Keith-Falconer. Born Mary Patricia Bell-Irving in 1925, she grew up in Vancouver's Shaughnessy neighbourhood. "My father was a graduate of Scotland's Loretto School. He returned to Vancouver in 1914 and went overseas with the Gordon Highlanders as a motorbike dispatch rider. He was wounded and spent thirty days unconscious in Lady Ridley's Hospital. Once he recovered, he joined the RAF, but remained a Gordon Highlander. As a result of his wounds, he wore a brace on his leg for the rest of his life. After the war he worked for the Gulf of Georgia Towing Company as a dispatcher and then founded Vancouver Shipyards. He then went into the flower business and finally insurance, an offshoot of H. Bell-Irving Company. He organized the air cadet movement in Canada and was Commanding Officer of No. 11 Squadron."

"Was he ever in the fishing business?" I asked.

"No," she replied, "only indirectly with the insurance business. Their offices were in the same building on West Pender, opposite the Marine Building. But he did like to fish, fly fishing, he even tied his own flies."

"Do you remember your grandfather, Bell-Irving?"

"Yes, but I was very young when he died, just six. He came to a dance recital that I was in. He loved parties and performances of all kinds. He bought Pasley Island [off BC's Sunshine Coast] in 1909. He was crazy about the island. We spent our summers there. Every morning he'd take a brisk walk around the island, a twenty-five-minute trip."

"To stay in shape for hunting?"

"Yes, my father told me that grandfather wanted to be thought of as a big game hunter. He suspected that the Indian guides that he employed on his grizzly hunting expeditions

HOPEFUL

In Good Hope's first year of operation as a lodge, 1970, seventy-five years after the first salmon hit its docks, these five tyees tipped the scales at 242 pounds, about 48 pounds apiece. Photo courtesy of lodge worker Kate Wilson (seated, left).

probably steered him well clear of prime grizzly locations. When a famous British big game hunter told him that he'd never shot a grizzly bear, he felt better for not having shot one either." She paused and laughed.

"What's so funny?" I asked.

"I was just thinking about Dad's brother, my Uncle Aeneas. Everyone in the family claimed that he was spoiled, because he was the baby of the family. I don't know about that, but he did have a pet bear."

"A teddy bear, you mean?"

"No, a real black bear! This was early on when the family lived on Seaton Street in Vancouver."

We talked for many hours about a wide range of subjects—not all of them related to Bell-Irving—over the course of three or four visits. Trish was the keeper of many of the Bell-Irving family records and photo albums, including a complete photocopied set of her grandfather's business diaries and his personal photo album. One afternoon as we were going through this album I was stunned to see three expertly shot photos. I recognized them immediately. These were the photos that Bell-Irving had taken of Good Hope just after it was built, mentioned in his 1895 diary. One was taken from the knoll to the south of the cannery looking north to the end of the cove. A second was taken from the end of the cove looking south across the fleet of skiffs to the cannery. A third captured the cannery looking directly at it from the opposite shore, the bow of a sailing vessel named the *Sultan* peeking in from the photographer's left. They were beautifully shot by Bell-Irving, his painter's eye for composition evident. Engineer, tycoon, big game hunter, painter and photographer—was there anything he couldn't do?

"I could kiss you, Trish!" I blurted.

"Why, what did I do?" she asked.

"You've lovingly preserved this history, that's all! These are photos of Good Hope!"

"Oh, so that's Good Hope, I'd never seen it before."

HOPEFUL

~

In Good Hope's second year of operation as a fishing lodge, 1971, Suzie McArthur joined the staff as her friend Barb Sutcliffe's assistant in the kitchen.

As we landed at the dock, Barb Sutcliffe informed me as she climbed on and I climbed off that she'd quit. I had literally never even boiled an egg before and was suddenly promoted to breakfast and lunch chef. Fortunately, it was short-lived, as Barb returned, but a little embarrassing in the interim when paying guests popped into the kitchen one morning and caught me toasting frozen Eggo waffles for their breakfast.

Black Johnnie was one of the local fishing guides hired by the camp who took great glee in waking us all up every morning at 4:00 a.m., banging on our doors with the refrain, "Drop your cocks and grab your socks, there's daylight on the swamp." The hours were long, most of us starting at 4:00 a.m. and some not finishing until 10:00 p.m., but still we managed to party many nights, often rewarded with a bottle of Scotch hijacked from the dining room. On our very infrequent days off (maybe one day every two to three weeks), we would take the company tug over to Calvert Island, anchoring on one side and walking across to the glorious beach on the other. Our biggest entertainment in the evening was listening to the VHF radio when often loggers or fishermen were calling home to tell their wives they wouldn't be coming home as soon as planned. [There were] interesting responses from many of the wives whose biggest concern was whether a paycheque was at least coming home.

I was often struck by the formality of the local residents, for example, the Nygaards and the Bachens, two couples who had lived in the Inlet for thirty-plus years and still referred to each other as Mr. or Mrs. when meeting in the store.

One day, we were startled to hear someone yelling, "Fire, fire!" There appeared to be a fire in the generator room. We had been forewarned that if ever there was a fire, everyone should just run to the boats and get offshore as the place would go up like a tinderbox. I remember running back to our house, grabbing the only thing of value I could lay my hands on and, clutching the item under my arm, I ran to get on the tug. It wasn't until the fire was extinguished and the alarm was over, that I looked at what I carried out with me. It was a box of glass string beads that I had just ordered through the mail valued at about ten dollars.

After a long, hectic, busy season, everyone's nerves were a little frayed. I recall getting into a yelling match with the manager and quitting in a huff. That same day, as I'm plotting my departure, an unexpected seaplane arrived and Mother stepped out to surprise me for my twenty-first birthday. I had to quickly make amends with the manager so Mom could stay on for a few days and be oblivious to the goings on behind the scenes. After the last guests left, my boyfriend (later to become my husband) and I asked to stay on as caretakers for the winter. We were looking forward to the adventure of having the whole camp to ourselves and to experience a winter up there. Unfortunately, the boss had also agreed to let another couple stay on too. Given that the other guy was a real rough cut from Port Alberni and his new girlfriend was one of the Japanese girls that had been brought in by the management to waitress, who spoke no English, suddenly that winter of isolation seemed like it was going to be a winter of too close quarters, so we bowed out. Our last day was Hallowe'en and we flew out to Port Hardy, enduring the roughest flight imaginable. Once we landed, the pilot said to one of his colleagues on the tarmac, "We shouldn't have been flying today."

~

In March 1972 Ian Bell-Irving hired a manager for Good Hope. When I heard his name I did a mental double take and then marvelled at the serendipity or coincidence or whatever it is that governs these things. His name was Alfred Lauritzen. Except for the "z," he was the third in a line of Lauritsens to manage Good Hope going back to the 1920s and Levi Lauritsen's Uncle Victor. So far as he knew, however, he had no Norwegian blood. Alfred Lauritzen was born in Belgium. He went into the hotel management business and worked in southern France, Germany, England, Spain and in Lisbon for three years. He came to Canada to work as the assistant manager at the Harrison Hot Springs Hotel before joining the Georgian Towers Hotel in Vancouver and then CN Hotels' Jasper Park Lodge. Alfred was referred to Bell-Irving. He wasn't sure he wanted the job, but once he saw Good Hope he succumbed to the natural beauty of the inlet and signed on. Alfred's wife, Counsuelo, and daughters Susanne, Cathy, Sylvia and Marychello would accompany him for the season. Gordie Howe and his wife Colleen were among the guests. The Lauritzen girls picked wildflowers for Colleen and made them into bouquets.

Good Hope operated for just one season under Bell-Irving's ownership, at a loss, and financial troubles with his other business ventures forced him to sell. In December 1972 Lauritzen put together a business plan to take over the cannery and had discussions with potential financial partners, but nothing came of it.

Robert Henry and Jeff Crawford owned and operated Good Hope from 1973 to 1979. Crawford was a Vancouver restaurateur, Henry a pilot for Eastern Airlines. Harold Dawes did electrical work for Good Hope from 1968 to 1975. Apprenticed to be an electrician at the age of fourteen in Manchester, England, Dawes spent a couple of years as an electrician with the RAF, participating in the Berlin Air Lift at the end of World War II. He immigrated to Canada and settled in North Vancouver, working for many years as an electrician with the school board. Harold's thoughts about Good Hope run to "disaster after disaster." So far as he was concerned, something was always going terribly wrong. "What would happen is that in the winter, logs would get underneath the cannery and bang repeatedly against the pilings, knocking some of them out." He recalled the fire that burned down the buildings to the northeast of the cannery, destroying the diesel engine and power generator in 1972. "The young guys working at the cannery had put their wet laundry over the generator vents to dry. The whole thing went up in flames." Harold installed two new diesel generators, a smaller one for times when less electrical power was needed, and a larger one.

One April day Harold Dawes and his son arrived at Good Hope with Jeff Crawford and Robert Henry. The plane dropped them off and promptly took off again. The winter caretaker couple (and their German shepherd), who had already departed the cannery, had been supplied with enough food to last them through April. But the four men discovered that there was nothing left. Also, the radio telephone batteries were dead. They were stranded without food or means of communication. One of the owners remembered that the office safe contained tins of salmon. "So we opened the safe and sure enough there were dozens of full tins. The only problem was that they were canned in 1946." The salmon was almost thirty years old, but they were hungry and opened a tin. It looked and smelled okay and they ate it. "No one died, so that's what we lived on for the next few days until the plane arrived to take us home."

Dawes remembered Good Hope's wild cats, possibly descendants of Marea Abelson's cats, coming into the kitchen.

They weren't afraid of me and they were ready to fight if I tried to chase them away. They had an old tugboat up there that they used to pull a log boom across the cove in the winter and back again in the spring. That's all they ever used it for. One time the float plane we were waiting for was delayed so we decided to get the tug started and take it out into the inlet. The swells were pretty big, so we turned back. As we were coming into the dock I put the boat's engine into reverse. The water foamed up and then the engine screamed. I knew immediately what had happened, the propeller had come off. We hit the dock so hard that things hanging on the wall in the kitchen rattled, they told us later. What had happened was the kids who worked up there had removed the propeller doing some maintenance work on it and had forgotten to replace the cotter pin that held it on. We were lucky that it hadn't come off when we had it out in the inlet.

One day a sailboat came into the cove. They drew my attention to a cloud of smoke rising from somewhere in the general direction of Wadham's. I took a boat and went there to see what the problem was. I found half the dock in flames and the caretaker completely blotto. He couldn't even stand he was so drunk. I found a chainsaw and incredibly it actually worked. I cut off the burning half and it collapsed into the water. So they still had half a dock anyway.

I was working in the cannery when I heard a girl in the kitchen screaming. She had accidentally banged a full container of hot coffee against a pipe and it had broken, spilling hot coffee all over her lap and legs. I ran in, grabbed her, and we both went over the dock into the water. A float plane came and took her to hospital.

A baggage handler at Port Hardy airport was stealing salmon. I lost a whole fish myself. But just imagine those guys who came from the US, to get home and discover that their fish had been stolen; it wasn't good for business.

One time I got a call from Bob Henry to hurry and get up to the cannery, there was no power. Bob was in a panic when he met me at the dock, so I said calm down, go and have a coffee and I'll see you in a little while. I had a look at the generator. A wire had become disconnected, so I hooked it up and the power was on again, even before Henry could get his coffee.

In yet another weird and wonderful Good Hope coincidence, I was out one evening with Darion Jones, an old friend from our days together at the University of British Columbia. We were having a glass of wine when I launched into what I was doing for a book on a cannery called Good Hope.

"Good Hope? You mean the fishing lodge?" she asked.

"Yeah, do you know it?"

"Duh, I worked there! I was there in… I don't even remember the year… I think it was 1980."

"Who owned it at that time?" I asked.

"Three doctors in Bellevue, Washington. I never met them."

"Who ran the show?"

"A guy named Denny Bowen. Oh yeah, it was a quite a summer. It turned into a real gong show," she said, laughing at the memory. "Denny hired me. I started out as a kind of liaison person, looking after the guests when they got to Vancouver, ordering and picking up supplies, stuff like that. Then Denny needed me up at Good Hope, doing whatever needed doing. So up I went. The cook was a 300-pound Haida named George. He quit almost right away and a Japanese guy, Satoschi, who had been the sous-chef, took over. I don't remember all the guides and staff, but there was a guide named Jan Brouwer; Forrest and Nola were a couple, he was a guide and she worked in the kitchen; there was Steve, Kim, Jill…"

"What was the gong show?" I interrupted.

"There was quite a lot of drinking going on with some of the guides and staff. The guides would be so hungover in the morning that they'd open up their packages of herring bait and puke."

"In front of guests?"

"Oh yeah. Denny wasn't around much, so a lot of stuff went on that shouldn't have. Three guides got fed up and left, so Denny hired three replacements from Red Deer, Alberta, who had never guided on the coast. I knew more than they did, but Denny wouldn't let me try guiding. Apparently, it wasn't a woman's occupation. By early August I wanted out, but

Denny wouldn't let me leave. So one day I got on the radio phone and told my dad to send a plane up to get me. Denny overheard the conversation and was pissed off, but the plane arrived and I left. Then Denny convinced me to come back by offering me more money. So after a couple of weeks I went back up. Around the end of August or early September the lodge went bankrupt. The float planes stopped flying in until they got paid. We were stuck there, some of the staff had major cases of cabin fever going; they hadn't been out in months. I had to talk one guy out of committing suicide. The lodge turned into a Club Med for staff. People started drinking all the booze and eating all the food. One afternoon, things boiled over. Some people started throwing furniture and stuff through the windows of the bunkhouse, smashing things, ripping memorabilia off the walls. It was like a scene from *Lord of the Flies*. The rampage went on for about an hour."

"Where was Denny while all this was going on?"

"He was holed up somewhere with his revolver."

"And you?"

"Playing ping-pong with the guests, pretending none of it was happening. The RCMP arrived, talked, but no one was arrested, and eventually planes arrived to take everyone out."

Doug Richards brought Good Hope back from the brink in 1984. His wife, Arlene, was the engine that made his vision an operational reality. Photo courtesy of the Richards family.

Later on I came across something written by Rivers Inlet Hospital founder Dr. R.W. Large. "If all would combine and forbid the landing of liquor on the Inlet, it would be practically settled, and with better police patrol and swift justice, much can be done. Some tragedy may be necessary to rouse all to a sense of the peril, but at present the getting of a pack of salmon with as little friction as possible is the aim of most of those who, if united, could do much, and the effects of the drinking, while deplored, are by many looked upon as inevitable. At the close of the season, most of the Indians purchase liquor from the Chinese, and there is a grand finale. Then we have very little more trouble with liquor till the following season."

The work was non-stop, but the Richards family pulled together to keep Good Hope alive. Shown here, left to right: Arlene, Doug, Jennifer, Ted and Susan. Photo courtesy of the Richards family.

~

There were times when my approach to research wasn't much different from a fisherman making a set from his skiff, only my net was the internet. I was "fishing" one day when I came across a message posted by Susan Richards de Wit. In it she talked about her family owning and operating Good Hope in the 1980s, but there was no contact information. Some months

THE GOOD HOPE CANNERY

later her brother Ted responded to an ad I ran in the papers asking for Good Hope alumni to contact me. Ted Richards worked with his dad, Doug, at Big Spring Resort in Rivers Inlet until Doug bought Good Hope out of receivership in 1984.

At Good Hope, Ted was involved with every aspect of the business including guiding, trade shows, purchasing, repair work and smoking salmon. When "Hollywood" came calling, he was ready, appearing as the guide in the video *Giant Salmon of BC*. He also guided for *The Waltons'* TV show creator, Earl Hamner, the two collaborating on stories as they left Good Hope in thick fog bound for the mouth of the inlet. "Good Hope was [my father's] life

Doug Richards bought Good Hope out of receivership in 1984. Photo courtesy of the Richards family.

dream. He liked the outdoors and he liked talking to people and sharing stories, and he could enjoy the outdoors every day and the stories every night." Ted put me in touch with his sister Susan, who, like the rest of the family—Bob, Jim and younger sister Jennifer—worked at Good Hope. Being the family historian, Susan graciously provided me with more information on the Richards era.

Doug and Arlene Richards had Good Hope until 1990, or rather Good Hope had them. It couldn't have fallen into better hands. The couple were stewards of Good Hope, running the operation as a family business. Doug Richards worked and lived on the water all his

Aerial view of the Good Hope Cannery Lodge, 1980s. Photo courtesy of the Richards family.

Doug Richards and a silvery trophy. Photo courtesy of the Richards family.

life, beginning as a boy when he would go fishing up Indian Arm with his father. From 1961 to 1972 he owned Richards Marine Service. His association with Good Hope began in 1973–74 when he served as the lodge's consultant on marine engines and boats. He worked in the same capacity at Big Spring from 1981–83 before buying Good Hope.

"Doug bought the lodge when it was really not livable anymore," Arlene explained. "He fixed it up and brought it back to life to be used again as a guest resort. It was his idea and initiative to start the Rivers Inlet Lodge Owners Association. He had the other owners to Good Hope several times and wanted to create a connection and sharing amongst the owners. As for me, my favourite spot was sitting in the dining room looking up the inlet. The sky and the water changed colour constantly, very pretty. There were always one or two fishing boats anchoring. Beautiful."

With heavy hearts the Richards sold Good Hope in 1991. When Doug Richards died in 2007, Good Hope lost one of its most caring and passionate custodians.

"There were a few years when Good Hope was not in operation," Susan said. "The lodge was empty. We tried to put the history back into it—clean up the machines and see what we could get working. Dad really wanted the old boat that was up there to stay there and represent the original feel. Dad was a naturalist first and a fisherman second. He wanted Good Hope's surroundings to stay pristine. Logging companies made him offers, but he wouldn't touch it. Good Hope was his passion, he lived for that place, he was at peace there. He was mechanical at heart, and loved Native ways. Dad was the entrepreneur and Mom was the engine. She did the work in the background, kept things going, did the human resources work, often the cooking and managing the running, greeting guests from overseas at the airport, bookkeeping, whatever was needed."

For Susan Richards, one experience stands out in her mind. "Dad and I were going from Good Hope to Port Hardy. He asked me to steer the boat in the direction of a point and

The prospect of a huge tyee brings fishermen to Rivers Inlet. Here Ted Richards, shown on the cover of the *British Columbia Sport Fishing* magazine, celebrates landing the big one. Image courtesy of Richards family.

went below to catch a little sleep. A big storm came up and I was pretty scared, but a huge blue whale surfaced, its eye seemed as big as the boat. Somehow it made me feel at peace. It stayed with me all the way to the point Dad had told me to steer toward."

A group of investors including Bob Stewart and Ivan Berry of Berry's Bait & Tackle in Richmond, BC, took over Good Hope in 1993 and continued to operate it as a fishing resort. Throughout the 1990s Kim and Tony Allard were frequent guests. Their first visit was for a couples' fishing derby, a charity fundraiser. They kept returning, eventually booking out Good Hope for a four-day trip each summer exclusively for family and friends. "It's close to Vancouver, but it feels remote, like going back in time," observed Tony. "We thought it was a marvellous place to hang out as a family. No TVs or cell phones and the nearest neighbours are miles away." The Allards, with their two young sons, Gage and Luke, had been looking for a summer place. "Growing up on the Prairies we'd spent lots of time with friends at their cottage on the lake. Kim and I wanted to find the equivalent on the west coast." Tony had gotten to know Bob Stewart and mentioned to him that if he were ever considering selling Good Hope he'd be a possible buyer. In 2006 they did the deal. The Allards had no master plan. "When we bought it we initially set out to just make the century-old building as safe as possible. One day we could feel the whole place shaking as large storm waves struck the pilings. So we put in three big steel piles by the gut house with a log boom in front to act as a shock absorber. And with the help and advice of logger and neighbour John Salo we just went from there, from the bottom up, starting with replacing piles and cap beams, rebuilding the gables, the net loft, putting in a fire escape from the second floor, constructing a new staff building [to replace the "Hilton"], a modern, environmentally safe sewage system, et cetera." The old network of walkways, decks and docks was replaced by a new and expanded system, including scenic viewpoints with seating and a covered arrivals/departures waiting area. In 2008–9 Good Hope's kitchen and staff dining areas were reconfigured and the guest rooms and bathrooms extensively renovated.

Frustrated with guests having to fly first to Port Hardy where fog too frequently necessitated delays in getting to Good Hope, the Allards decided on direct flights using the safe and reliable Cessna Amphibian Caravan float planes. This decision, in turn, meant that they would have to install a new fuelling system—as far from the lodge as possible. And so "Function Junction" came into existence, safely supplying not only gas but jet fuel, propane and diesel. "One day," said Tony, "we'd like to install a 'run of river' generating system for our electrical needs, thereby further reducing stress on the environment.

"Over the years we've developed great affection for Good Hope and the guests, guides and the rest of the staff who keep coming back. It's become an extended family. An unanticipated joy for us has been getting to know the people of the inlet, the Oowekeeno, and their thousands of years of history here. Our boys have even been ceremonially adopted into the Walkus family in a potlatch ceremony, a very special honour. The people of Rivers Inlet, the Oowekeeno, have a lot to teach us about family life and many other things as well. I am just upset with myself for taking so long to get to know them, so I will try to make up for lost time."

The *Silver Queen*, Good Hope, 1956. Photo courtesy of Gerry and Rosemary Miller.

Chapter 10

Self-propelled Canoes

In our culture a chief is not a chief until the people tell him he is. He has to earn it, he can't just assume it.

—Charlie Willie

As the twin 150-hp Yamaha engines hurtled our aluminum-hulled Ironwood at 31 knots over the stone-smooth waters of Darby Channel at 6:10 a.m., 27-year-old guide Aaron Beeching cranked up the boat's satellite-radio-fed sound system—"Ramble On" from *Led Zeppelin II*. Low slung against the far shore and outlined with bright white connect-the-dots Christmas lights, Dawson's Landing appeared and disappeared to starboard like a saltwater mirage. *Ramble on, ramble on.* I poured myself a cup of fresh, hot coffee from what was likely the world's best thermos. It was late August 2009 and I had come to Good Hope for a second time. Marea Larsen was up again, too, along with her son Carl and her 15-year-old grandson (and my nephew), Ryan MacDonald. Marea was not a morning fisherman, but Ryan and Carl were pinned by velocity to the aft bench seat, arms crossed, their faces buried in drawstringed hoodies. We were headed west to Calvert Island in Fitz Hugh Sound to fish for coho in a place they called the "golf course" because of some un-forested swathes visible from the fishing grounds.

Back in the old days they'd be picking up their nets about now, I mused, working out the kinks in their backs after catching a few hours of sleep on board their oar- and sail-powered wooden skiffs. About the only thing their boats and ours had in common was the length, 25 feet [7.5 metres]. I could imagine them well enough, with the advantage of hindsight that time bestows on the present, hundreds of skiffs adrift on the tide, but could they imagine us? And what would they think? What would Led Zeppelin sound like to them? These were men of the music hall and the phonograph, of Tin Pan Alley tunes and early radio featuring the Paul Whiteman Orchestra. They'd never heard blues music, let alone rock 'n' roll.

I snapped out of my reverie. Aaron was all business, eyes straight ahead. He wasn't imagining old-time fishermen, but he may have had mattresses on his mind. The day before, while we waited for a strike, Aaron turned to me and said, "They closed the Sears in Massett, so now I'll have to drive to Skidegate to pick up this mattress that we ordered. How am I going to get a mattress from Skidegate to Massett?"

"How about a truck?"

"I don't have a truck."

"How about a van?"

"No van either. Maybe I could strap it to the roof of my car, but the rain…"

At the "golf course" Aaron killed the Yamahas in favour of the small trolling motor. In minutes he had four hooks baited and lines twenty-four pulls into the water. He cleaned his knife, checked all the lines and poured himself a tea. "There's something about the tea they make at the lodge," he said. "It's just Earl Grey, but it's really good."

It is wonderful how quickly time passes while you are waiting for a salmon to bite. If you were doing anything else it would be an intolerable length of time. When it is a salmon you are waiting for, somehow it is more than all right, it's just fine; you are content beyond waiting. Trolling off Calvert Island, there is everything to look at and nothing to see, the world is elemental and vast, boats come and boats go, rods bend and go slack again, the sun shines or it doesn't, whatever is happening is happening somewhere in the opaque jade depths of the ocean.

Zing! Ryan, half-asleep, was now very awake. He grabbed the rod and set the hook. "Reel! Reel! Reel!" yelled Aaron. "Tip up! That's it! Let him go… okay, reel! That's it! Tip up!"

SELF-PROPELLED CANOES

Carl, Aaron and I reeled in the other lines. In a few minutes the coho broke the water, flashing in the sun like medieval armour.

"That's it, Ryan, that's it, good job, but keep your tip up."

The coho made one last break for it. *Zing!* He was running! Ryan let him go and then the line slackened. Ryan reeled… and reeled… and reeled. The coho was just 10 feet (3.04 metres) from the boat, just a couple of feet beneath the surface. Ryan reeled again, tip up this time. The coho darted to the stern.

Marea Larsen and grandson Ryan MacDonald, Rivers Inlet, 2009. Photo by Carl Larsen.

"Don't let him go where he wants to go," said Aaron. "You've got to lead him where you want him to go."

Ryan, a quick learner, led him back to starboard. Carl readied the net. One last attempt to free himself, one last flip, and Carl had him. *Whonk!* Aaron delivered the coup de grâce to the top of the fish's head, removed the hook, slit the coho's gills to bleed him, and slid him into the holding tank. That's one. The limit is four per person per day. There was no limit in the days of the canneries. I thought of a line from an 1873 article in the *British Colonist* cleverly summarizing the business of canning fish: "Catch all they can and can all they catch." I thought of that day in July 1896 when 43,000 sockeye were delivered to Good Hope. Later, I would come across an even earlier piece from 1867 in the *New York Times*:

> The fisheries of the whole Northwest Coast have, however, never been touched with the magic wand of capital, but the day is not far distant when they will afford profitable employment to a large number of mankind. As a source of future wealth, the fisheries are infinitely superior to the fur trade. Trappers may disappear and the supply of skins will fall off, but the bountiful hand of nature may be depended upon for *a never-failing supply of the inhabitants of the deep* [emphasis mine]. In short, the fisheries may be set down as the most important branch of industry that will flourish on the coast.

Herring baited, the lines went back out, another twenty-four pulls. Aaron splashed a bucket of sea water over the deck, rinsing salmon blood into the ocean. Waiting for the next strike, we watched two humpback whales, tight against Calvert's rocky shoreline, doing some fishing of their own. "They're bubble netting," said Aaron. A group of humpbacks swims in a shrinking circle blowing bubbles below a school of prey. The shrinking ring of bubbles encircles the school and confines it in a tightening cylinder. The whales then suddenly swim upward through the bubble net, mouths open wide, swallowing thousands of fish. The ring can begin at up to 98 feet (30 metres) in diameter with the co-operation of a dozen humpbacks. Some of the whales blow the bubbles, some dive deeper to drive fish toward the surface, and others herd other fish into the net by vocalizing.

SELF-PROPELLED CANOES

Humpback whales are mammals like humans. They are warm-blooded, breathe air, and they bear live young, nursing them with milk. The humpback has a distinctive body shape, with long pectoral fins and a knobbed head. Black with white patches on the flippers, bottom surface of the tail flukes and body, humpback adults range in length from 39 to 52 feet (12–16 metres) and weigh almost 40 tons. They are acrobatic animals that breach repeatedly and have a spectacular blow of spray. Males produce a complex whale song, which lasts for ten to twenty minutes and is repeated for hours at a time. The humpback is usually found in small groups but groups as large as two hundred have been known. Found in oceans and seas around the world, humpback whales typically migrate up to 15,500 miles (25,000 kilometres) each year. They feed only in summer, in polar waters, and migrate to tropical or subtropical waters to breed and give birth in the winter. They mate every two or more years and after a year-long gestation period a 15-foot (4.5-metre), 2-ton calf is born. During the winter, humpbacks fast and live off their fat reserves. Their diet consists mostly of krill and small schooling fish such as herring, salmon, capelin and sand lance. Humpback summering grounds extend from British Columbia to much of the Gulf of Alaska. They winter off Baja, California, and Hawaii. Fewer than two thousand humpback whales exist in the northern Pacific, although the numbers are slowly increasing.

By late morning our fishing methods had yielded us eight coho. Aaron fired up the big engines and we headed back to Good Hope.

~

Inside the cannery sits the gillnetter *NP 173*. The "NP" designates it as having done service at North Pacific, an ABC company cannery on the Skeena River. When ABC boats had been worn out by the tough conditions on the Skeena, they were sent down to Good Hope where the relatively calmer weather of Rivers Inlet meant that, with the help of skilled mechanics, such as Cecil Fisher and Gerry Miller, many more years of service could be wrung out of the old boats. Marea Larson identified *NP 173* as having belonged to Good Hope net man Paul Gertner. "Paul Gertner and Pete Pederson were friends and neighbours in Aldergrove. Both

THE GOOD HOPE CANNERY

Net men, Good Hope, 1948. Photo courtesy of Rolf Hundvik.

were bachelors. Pederson had had a fishing accident and wore an artificial leg, which I think must have been very painful. He stayed upstairs in the cookhouse. It was difficult for him to get up and down those stairs, but he never complained. He was a very nice man." Seeing the boat awaiting some tender loving care, I was reminded again of Hagar Shipley in *The Stone Angel*, who describes a similar old boat in a cannery: "At the far end of the long room stands a derelict fishboat, perched up on blocks, stripped of gear and tackle, faded blue shavings of paint falling away from its hull. Not even a ghost vessel, this. Only a skeleton, such as one that might have been washed up somewhere centuries after it had set out for heaven with its Viking dead."

Outside the dining room Native carver Charlie Willie was working on his Sisiutl carving. On my first day up he'd told me about it. "The Sisiutl is the two-headed sea serpent that guards the entrance to the homes of the supernaturals. Sisiutl was believed to kill and eat anyone who saw it, kind of like Medusa. They say that washing in its blood turned a person to stone and that it could be transformed into a self-propelled canoe that must be fed seals."

Self-propelled canoe? "I was just out in a self-propelled canoe," I said.

"Does it run on seals?" he asked, laughing.

Whether carved or painted, the Sisiutl is depicted with a profile head, teeth and a large curled tongue at each end of its serpentine form. In the centre is a human head. Fins run

192

along its back and curled appendages or horns rise from all three heads. The painted body represents scales and it may be carved horizontally, formed into a U-shape or coiled into a circle. "They'd paint Sisiutl on the sides of canoes and over doorways to protect the people from evil spirits," Charlie explained.

Later that afternoon he introduced me to his father, Paul. The three of us talked, leaning on the railing by the gut shed. Paul was born in Kingcome in 1940, a village made famous by Margaret Craven's 1967 novel *I Heard the Owl Call My Name*. He fished for Good Hope with his own boat, the *MM*, from 1954 to 1962. "A boat was cheap in those days," he told me. "I paid $1,500, the engine was a one-lunger Easthope. They were paying 38 cents a pound for sockeye back then. A small catch for me was a hundred and fifty a day. My biggest catch was about four hundred in one day. You could do pretty well, but I was young and spent it all," he said, laughing.

Paul recounted his most vivid memory of fishing: "One midnight in the inlet a pod of killer whales went after the salmon in our nets. Our nets were about twenty feet apart and we had to pull them in."

Back in the present, Charlie carved off a sliver of wood and said, "By the way, I talked to my dad about that fisherman." I'd brought up the subject of a fisherman I'd been told had drowned in the inlet. He had fallen overboard as he and his five-year-old son were going out fishing. Struggling to stay afloat, he yelled at his son to kill the engine, but his son didn't know how. The boat kept going and the fisherman drowned. "His name was Paddy Henry," Charlie told me. "And he was a first cousin to my Dad." Was there no end to this string of Good Hope coincidences? "The story is just like you heard it. His son's name was George, Georgie they called him. They say Georgie never recovered from his father drowning that day. He died young." After a pause, he added: "My dad also told me about another fisherman who drowned here in the inlet, Henry Nelson. They didn't find him until a day or two later. His son Hank was fishing and his dad's body came up in his net."

Good Hope has had its share of unfortunate and unsettling deaths throughout the years, among them Chief Jumbo's infant daughter in 1915, Werner Smedman and Thor Ovitslund

THE GOOD HOPE CANNERY

in the 1940s, and Dick Haggerty and Paddy Henry in the early 1960s. Are their spirits haunting Good Hope? "Any guests who stayed in Room 9 would get an uncomfortable feeling, enough to ask to be moved," Susan Richards de Wit told me. Ian Bell-Irving reported boat horns going off for no apparent reason in the middle of the night. At dinner I asked our server, Laura, if she'd experienced anything ghostly at Good Hope. "I think most of us have. I know that a toothbrush of mine went missing, just vanished, not lost, not misplaced, just gone. Who steals a toothbrush?" Another female staffer reported waking up terrified to dis-

Medical and missionary services in Rivers Inlet were pioneered by Dr. R.W. Large and later performed by Dr. George Darby. Pictured is the Rivers Inlet hospital and mission, c. 1906. Photo courtesy of the United Church of Canada.

cover what appeared to be a tall man pinning her immobile to the bed. The apparition vanished. Guide Aaron Beeching, hardly an excitable type, wasn't the only one to report a number of instances of seeing or sensing someone else in a room and of inexplicable blood spots in a sink. Chef Robert Hedley told me about a number of "hair-raising" experiences in the kitchen. "I had just turned away from stacking boxes of pasta on a shelf when suddenly they all fell over like dominoes. For some reason it caused the hair on the back of my neck to bristle. It struck me as weird because everything was level, but I straightened them up again and turned away. Well, they fell over again. I don't know what made me do it, but I swore at the 'ghost' and haven't had any weirdness since."

On my first trip up I hadn't been able to find any sign of a graveyard that existed somewhere on the north side of the inlet near Sandell Bay. Marea Larsen and I had spent a couple of hours looking for a spot I'd read about in *Drums and Scalpel*, an account by Dr. R. Geddes Large, published in 1968, of life in Rivers Inlet as experienced by him and his family. His father, Dr. R.W. Large, was a United Church missionary and doctor practising medicine and ministering to the "man-eating savages" just after the turn of the century. In 1904 a fire had destroyed the closed Wannock Cannery, taking the original Rivers Inlet with it. Green's Cannery, across the bay from Wannock, had also closed down and was used as storage for nearby Brunswick Cannery. Large explains, "Here [at Green's] the hospital was operated the following year [1905] using the old cannery buildings. In the fall and winter of 1905–06 the new hospital buildings were erected. They were built on a point of land that formed the cove in which the cannery had been placed, and the ground of the point, being partially cleared, formed a natural park."

Cyril Douglas had landed me at what I thought was the site of Green's, but was actually Brunswick. There were bricks scattered among stones on the beach, and, submerged under trees and scrub, a 4-by-6-foot [1.2 x 2 metre] square brick and concrete structure. My guess at the time, due to its proximity to the water, was that it had been part of a wharf. This conjecture was confirmed by Large. "In 1909 the provincial government built a board sidewalk to replace the trail so that hospital supplies and patients could be more easily transported between the hospital and the wharf at Brunswick." After happily wandering around the site aimlessly and finding no trace of a graveyard, I rejoined Cyril and Marea on board the boat.

I was no better prepared to find the graveyard on my return trip. In fact, recruiting Carl and Ryan, we set off to find it by going back to the Brunswick site, which I still believed to be the site of Green's. On the way there, Aaron remembered seeing the graveyard ten or more years earlier, but couldn't be certain where it was. While he manned the boat, we ventured farther up the mountain in behind the Brunswick site and found numerous piles of old bottles, cracked wash basins and rusted pieces of steel half-buried in the ground. But no graveyard. I decided, on no strong evidence, that we must be looking in the wrong place and we re-boarded the boat. The ever-patient Aaron dropped us off again approximately halfway between Brunswick and Green's.

"Let's see if we can find the boardwalk that used to run between the two canneries," I said.

"How long ago was this?" Carl asked.

"About sixty years ago, give or take a decade," I replied.

"Do you really think it could have survived this long?"

"No. I'm just trying to be encouraging."

We set off. An hour later, drenched with sweat, having caromed off stumps, pushed through alder even a machete couldn't have cleared, and crashed through salal-camouflaged trap doors, we hadn't found the slightest trace of a boardwalk, a graveyard, or anything else remotely human in construction. In fact, it was impossible that anything man-made had ever stood in these woods.

"Where are we, anyway?" asked Ryan.

"We're lost," I replied. "Where the hell is Carl?"

"He was right behind me about ten minutes ago."

"This is poor trek leadership skills, I may have to fire myself."

In a few minutes Carl arrived.

"What kept you?" I asked.

"I was filming."

Carl was a professional TV cameraman, armed today with a small digital recorder.

"Right. Did you happen to see a sign saying 'Boardwalk this way' or 'Graveyard dead ahead'? Anything like that?"

"Not a thing. What do we do now?"

"Retreat. Sound the bugle."

We boarded the boat and this time Aaron dropped us off in the small cove that had once been occupied by Green's but which I still believed was Brunswick.

"I remember a big dozer in the woods. It was somewhere near the graveyard."

We set off. Sometime in the last twenty or thirty years a logging company had used the site as a staging ground. They had constructed a landing, now home to a thick forest of alder. To one side we found a pile of rusting rollers, looking like unexploded shells. Farther up we discovered a boom boat wedged into a ravine. How it came to reside there was a mystery. We climbed a hill, searching for a road, a trail, a dozer, anything at all. We went in a circle. Twice. Found nothing but slash, stumps, salal and mosquitoes. From the boat we'd seen a cabin on a point of land. Was this Large's point, the "natural park" that he'd written of? That would be too obvious and easy, but we were in the mood for obvious and easy. The wooden stairs to the cabin's porch had rotted away, but someone had substituted a ladder. The porch itself was teetering on the brink of separating from the cabin. "Watch your step," I warned the guys as I navigated a particularly rotten section of porch. Opening a screen door, I stepped into the mud room. A blue plastic drum was full to the brim with fetid rainwater, the boards beneath it sagging with the weight. The inner room contained two cots, a table and chairs, a sink, cupboards and a cast iron oven. At the foot of the bed, a pair of shoes. On the table, an empty package of cigarettes and a cribbage board. On the windowsill above the sink, a teacup holding a toothbrush. On a shelf beside the sink, a 20-ounce can of beef stew. It was as if the occupant had recently left, if twenty years counted as recent.

"Needs some TLC," said Ryan.

"More like TNT," said Carl.

"Well, we've found a graveyard," I said. "Just not the right one."

Having exited the cabin, we explored the ruins to its immediate northeast. One or more wood buildings had once stood where we were standing. They had all long since collapsed and lay flattened out, a sea of weathered grey boards. Here the white enamel of half a sink floated on the sea, there a fireplace grate bobbed along. Had this been the place that Large was thinking of when he wrote, "From our home on the point we had a beautiful view up the Inlet and on Sunday afternoon we could watch the long tows of boats behind the cannery tug proceeding to the fishing grounds for the opening of the week's fishing." A few hundred feet away, perched out over a sheer drop to the jagged beach rocks below, a rather attractive privy stood, looking very sturdy and serviceable, in no danger whatsoever of collapsing or tumbling down. It made sense to build your privy to the highest possible standards, sparing no expense. After all, who wanted to die from an unfortunate fall while evacuating his bowels? Not I, and certainly not Dr. Large. At any rate, once again we had failed to find the graveyard.

The following day, having consulted Gordon Baron, a Rivers Inlet resident, watchman and photographer, we set off once more. Tony Allard volunteered to captain the boat and Marea to keep him company while Chef Robert, Carl, Ryan and I attempted to find the elusive burial ground. Gordon had assured us that it was located on the western side of the cove, to the right of a stream, beside an abandoned trailer, just a couple of hundred feet up from the beach. How could we miss? Once more our friends the alders did their best to block our progress. It was enough to make you wish that the Cannibal of the North had captured the four brothers and devoured them. We found the stream and the trailer, another graveyard in its own right, and scoured the area for what we had come for. We found a logging road and I followed it while the others broke off in a sort of flanking manoeuvre. Crossing a log bridge, I found the huge dozer that Aaron had remembered, parked at the dead end of the road. Was there nothing that a logging company wouldn't ditch in the bush after they'd mowed down all the trees? Trailers, boom boats, bulldozers, what else?

"Hey! We've found the graveyard!" yelled Ryan.

They were just south of where they'd flanked off. But before I could reach them I had the answer to my question of what else a logging company could ditch. There, parked at the entrance to the graveyard, was a worker transportation vehicle they called a "crummy" because it was. Spare parts and junk littered the area around it. Inside was a filthy mess of grease, rags, tools and thousands of nuts, bolts and washers. The graffiti jokes on the walls were all variations on the theme of Terry Fox and his 1980 Marathon of Hope cross-Canada run to raise cancer research funds. The graffiti dated the loggers' presence here to that summer.

And what of earlier summers? Whose presence then? Not loggers. Here had stood, I discovered later, Rivers Inlet Hospital, an attractive, well-built, two-and-a-half-storey building with a balcony overlooking the inlet. Using it as his summer base, Dr. George Darby served the medical needs of Rivers Inlet's population from 1914 to 1959. The one-room shack we had gone into likely stood on the site of the original manager's house, the one Darby had toured with his bride, Edna, in August 1914. The sea of grey boards was without doubt the "private park" he had, according to his biographer Hugh McKervill, enthused about, "a nice place for the children to play—when they come along." To which his wife had responded, "So long as they don't fall off the edge." And it was here that later in August the Darbys learned that Britain had declared war on Germany. He considered enlisting, but his wife convinced him that his medical skills were just as badly needed in the inlet and at the year-round hospital in Bella Bella. The Darbys had spent fifty-three summers here, but float planes and fast boats rendered the hospital obsolete. It was closed in 1957, a first aid station at Wadham's manned by a medical student taking its place.

At one time the cemetery must have been a peaceful site—on a relatively flat, open piece of ground, beside a stream, on a rise overlooking the bay. Over time the forest had grown back in and the ground appeared sunken. Trees, alder and scrub blocked the view, not that any of the residents were complaining. It was a dark, dank, moss- and fern-filled place now, the stereotype of the gloomy graveyard. And the trailer and the crummy? Degradation. It wasn't respect that had saved the gravestones; it was only superstition.

There were nine memorials in total, seven in English and two in Chinese. Moses Paddy "Nahsahuck" of Joheleset West Coast Village died at Rivers Inlet, June 25, 1910, aged 10 years.

THE GOOD HOPE CANNERY

Alexander Brown died August 26, 1907, aged 64 years. Jan Hervé "Frenchy" died suddenly in his boat at Rivers Inlet, August 1907. Edward Donnie, son of Alex and Daisy Mearns, was born January 21, 1919, and died August 16, 1919. John Morrison died at Rivers Inlet August 14, 1907, aged 30 years. K. Takahashi drowned July 8, 1907, aged 25 years. The daughter of Chief John Jumbo of Nootka, Christine, was born January 9, 1915, and died at Good Hope Cannery, July 23, 1915.

The year 1907 had been deadly: Brown, Hervé, Morrison and Takahashi. One died on his

Logging company "crummy" and assorted junk at the site of the Rivers Inlet Hospital cemetery. Photo by Carl Larsen, 2009.

boat, one died in the water. One was ten years old; two were infants, one a boy, one a girl. The girl had died at Good Hope. Beyond these bare facts, silence. Green's Cannery was defunct before the hospital was built. The hospital closed forever in 1957. Brunswick Cannery, a quarter of a mile away by boardwalk, had been built in 1897, two years after Good Hope, by George Dawson, Alfred Buttimer and George Wilson. Robert Jamieson had worked for them from 1899 to 1907. Had Jamieson known Brown, "Frenchy," Morrison and Takahashi? What did he know about the deaths in 1907? Was it just a coincidence that it was his last year

As of 2009 the tombstones were holding up remarkably well, in contrast to all but two of the inlet's canneries. Photo by Carl Larsen.

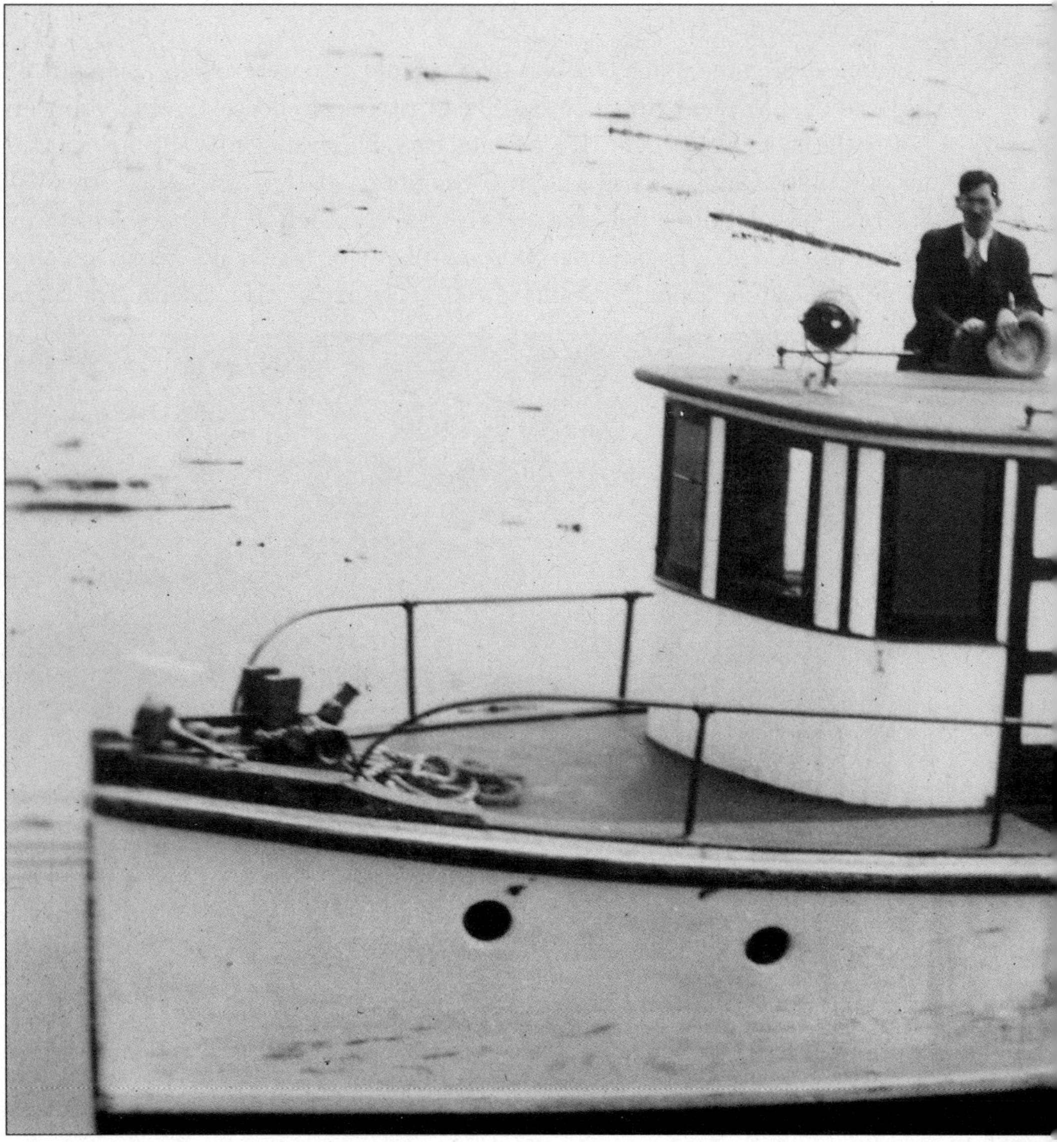

Dr. Darby aboard his marine ambulance in 1928. Photo courtesy of the United Church of Canada.

at the cannery? Brunswick was demolished in 1945, the last year of the Second World War, the last season of the three seasons captured in the bookkeeping binders that came out of the wall at Good Hope. Marea Larsen was ten. Levi Lauritsen ruled the roost. Decades later the loggers came and went. Their junk aside, only the gravestones remained. And, of course, Marea, who, along with Tony, had been patiently waiting for us on the boat.

"Did you find what you were looking for?" she asked, smiling pleasantly, when we returned.

"Yes," I replied, "and then some."

But I wasn't finished with graves yet. There were reports of a memorial on Ida, the first island out of Good Hope, only a couple of hundred yards to the southwest. "You can see the marker, or at least you used to be able to see it a few years ago," Cyril told me.

One evening before dinner Carl and I hauled two kayaks from the cannery to the water. A few minutes later we were ashore on Ida and looking for the grave. True to form, I was up to my waist in salal and heading in the wrong direction when Carl yelled out, "I've found it!" The location was only a few hundred feet from where we'd beached the kayaks. The plywood marker had been nailed to a tree and braced with two by fours, but nature had knocked it to the ground. As well, even if it had still been standing, it wouldn't have been visible from the water: a large, weathered root ball sat in front of it on the beach like a giant carved octopus. The marker was in three pieces. One piece read: "In Loving Memory of Paddy Henry—Age 37—Drowned July 27, 1960." A second piece featured a carved wooden sun symbol, a painted face in the centre with seven carved rays radiating out from it. The third piece had been cut in the shape of a humpback whale from a single sheet of plywood and outlined in black paint. The three pieces at one time had formed a single memorial. We carefully stacked the pieces in an upright position against the base of the tree.

Walking back to the kayaks we spotted a seal in shallow water close to shore. As we watched, it became clear that he had been hunting a fish, finally trapping it against the rocks. The fish, now clenched in the seal's mouth, flapped and splashed, and then the seal slipped silently beneath the water. It was the first seal we'd seen on the trip. Strange, I thought, that the first one we see is at the site of Paddy Henry's memorial, and stranger still that we were privileged to witness its fishing skill. Pondering this, a thought surfaced: the fisherman sinks beneath the water, returning to the domain of the fish.

It was no different with the Anglo-British Columbia Packing Company. When ABC began operations in 1891 its share of the total BC salmon market was 25 percent. Despite building Good Hope and acquiring other canneries over the years, ABC steadily lost its commanding position. By 1924 the company represented only 9 percent of the industry in the province. The agency agreement that Bell-Irving had secured at the outset had been so profitable that the company was opposed to ABC amalgamating with others. Any such deal would have cut off their agency profits. From 1921 onward ABC's losses exceeded half a million dollars. The company's shareholders were unhappy with the situation. On the other hand, since 1891 ABC had paid out in dividends to its shareholders two million dollars, or about four times the amount of their original investment. In 1925 Bell-Irving had made the difficult decision of passing on the presidency of H. Bell-Irving & Company to his second-born son, Richard, over first-born Henry. Henry Sr. and Jr. had a long history of skirmishing, but the precipitating incident was Henry Jr. ordering a packer boat built without first obtaining the permission of the ABC board of directors. Henry Sr. was incensed. Within the family, accusations and counter-accusations rang back and forth like a Shakespearean drama. As first-born son, Henry could expect to inherit the company. But as guardian of the interests of the ABC Company, Henry Sr.'s first duty was to the shareholders "who were good enough thirty five years ago to entrust me with their money." He stood by his decision and Richard took charge. Richard had been a lieutenant colonel in the Royal Flying Corps in World War I. Well connected in Great Britain, he was instrumental in the creation of the British Preferential Tariff instituted in 1932, giving Canadian canned salmon an advantage in British Commonwealth countries. Henry Jr. resigned in 1928 to join ABC rival Canadian Fish Company. Henry Ogle Bell-Irving died in 1931 at age 74. His estate ranked him sixteenth among Vancouver's elite with net assets of $339,000, but his generosity within his extensive circle of family and friends as well as his financial contributions to a diversity of public initiatives and causes was of an even higher rank.

On our last night at Good Hope, after one of chef Robert Hedley's five-star meals, after

Cyril's congratulations and awards to the people who'd helped catch a record number of chinooks that day at the head, after a moving speech by guest and renowned Native artist Roy Vickers, Charlie Willie took the floor as he had every night of our stay, and talked in his strong but modest way about the virtue of people not taking anything for granted. "In our culture," he said, "a chief is not a chief until the people tell him he is. He has to earn it, he can't just assume it." He then beat his drum and sang his most beautiful song yet, the story of what took place after the Great Flood.

> Kawadelekala, the wolf, was the Creator of my people at Kingcome Inlet. He wanted to see if there were any survivors of the flood. He howled to the North and nothing, howled to the East and nothing, howled to the South and nothing. When he howled to the West he got an answer. A wolf howled back from the Tl'atl'asekwala [Hope Island]. After the Great Flood, people on canoes were pulled away by currents of the flood and created new tribes where they landed. This is what is believed to have happened to the 'Wuikinuxv, that two of their canoes that got carried away by the flood waters created two other tribes, one to the north and one to the south of Rivers Inlet.

Charlie's stories, like Simon Walkus's, were a revelation: they made perfect sense. The transformations described in them made me think of other transformations, of an inlet associated for millennia with the Oowekeeno becoming in an instant associated with an English aristocrat who never set eyes on it, of cedar bark becoming Irish linen, of canoes becoming skiffs and human lungs becoming "one-lungers," of knife-wielding Chinese men becoming mechanized Iron Chinks, of pounds of silver fish becoming pounds sterling. Was it just change? Or progress? Or innovation? Or was it something else, something more like magic? Something more like the stories told by Walkus and Willie?

Later, after tasting Hans' "better-than-candy" smoked salmon in the upstairs lounge, I found my way to bed. As I lay there I thought back to the two long, black, aerodynamic birds—possibly Brandt's cormorants—that had seemingly raced our 300-hp boat down Darby Channel as we returned from an afternoon of fishing. They appeared out of nowhere, on our starboard. Aaron had the boat going nearly full throttle. The pair flew inches off the water, staying just a few feet ahead of us, until, for no apparent reason, they turned on whatever supernatural jets they possess, and zoomed effortlessly ahead of us, holding their rate of speed with utter ease, an estimated 60 knots. Quite frankly, they blew the doors off us. And then, when I thought they couldn't possibly go any faster, they swerved to their right and executed a long sweeping turn into a bay before coming back toward us. Considering the distance they'd travelled in sweeping through the bay, they ought to have come out somewhere behind us, even with their great speed, but instead they had actually increased their lead. They were so far ahead that I couldn't see where they came down, but I knew somehow that it must have been on the water opposite Dawson's Landing, now visible on our port side, its necklace of white Christmas lights just beginning to glow in the twilight.

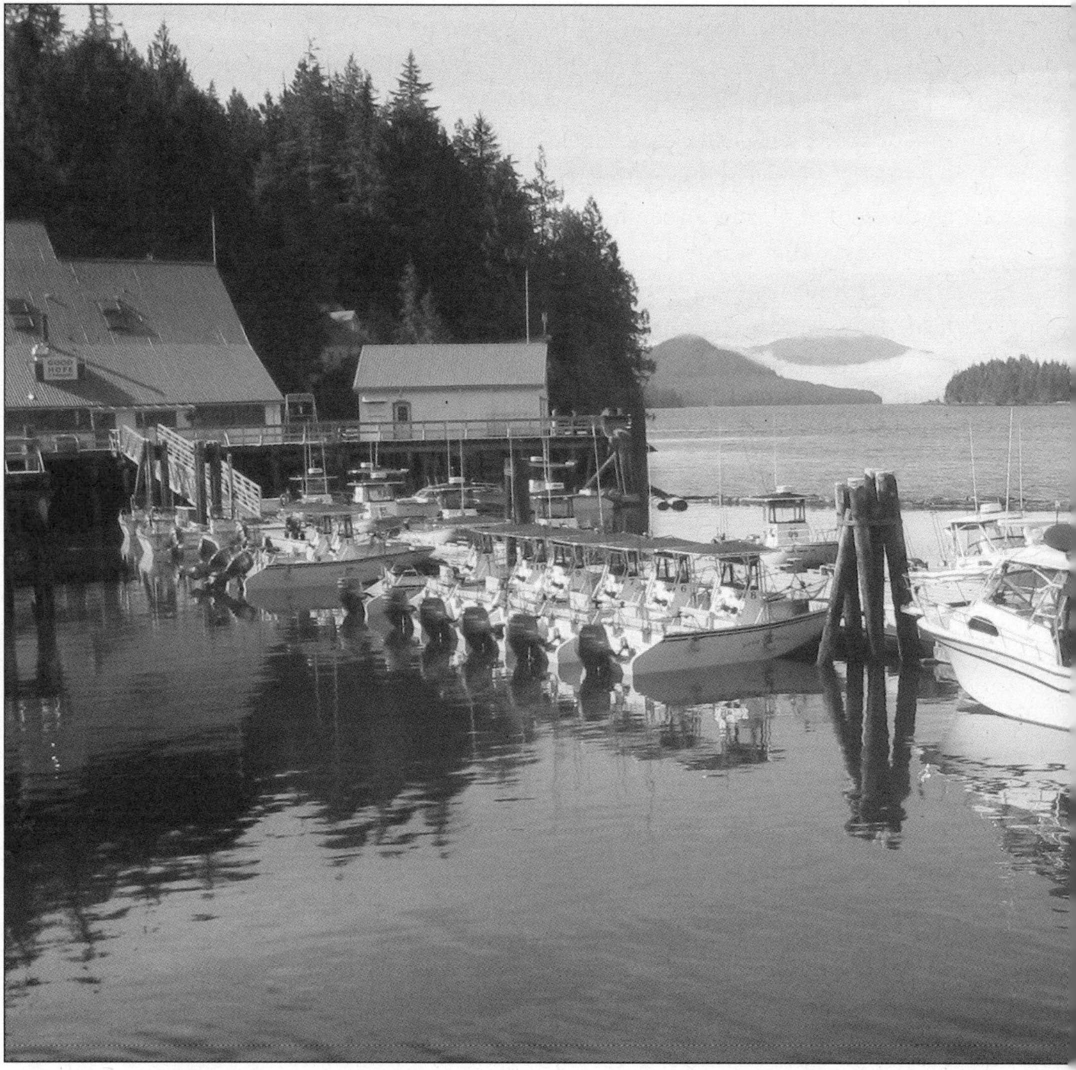

Tony Allard's vision has tranformed Good Hope into world-class sport fishing destination. Photo David Hayes, 2009.

Epilogue

As I wrote this book I entertained the idea that I was casting a net, catching in it pieces of history as if they were fish. But one night after listening to Charlie Willie, I began to think of my job differently. I wasn't catching anything. My role was on shore. If anything, I was a net man, tending linen, repairing it wherever it was torn, treating it in bluestone. I was copper sulphate, dissolving the algae of time. Whatever had been, was still there. There was nothing to catch; it had already been caught.

Acknowledgements

Without Tony Allard, there would be no book. His passion for Good Hope and his generous support over the two years it took to research and write this book made it possible.

I am indebted to Marea Larsen and her circle of friends from Good Hope: Gerry and Rosemary Miller, Mildred Dalton and Inga Fenwick. It is hard to imagine this book without them. I would also like to thank all the people who contributed a piece or two of the puzzle by sharing their Rivers Inlet/Good Hope memories, stories and photographs; and thanks to the staff of the City of Vancouver Archives and the University of British Columbia Rare Books and Special Collections for their help.

My sincere thanks to Ray Allegretto, Brian Nygaard, Patti Koenig, Janet Davidson, Cass Lindsay, Jane Yip (Uchida), Lorraine Behan, Chris Weicht, Tom Miyazaki, Frank Hanano, Jesse Rice, Gerrie Jackson (Fehr), Patrick Miller, Tina O'Connor, Jim Cox, Heidi Henderson, Randy Clark, Ron Clark, Pat Fiander, Bill Ridge, Ted Walkus, Russ Quinn, Arlene Richards, Susan Richards de Wit, Susanne Atkins, Marlene Yurichuk, Dick Simmonds, Derek Low, Rusty Harrison, Ron Finlayson, Lucky T. Bachen, Robert and Nancy Critchley, Paul Willie, Charlie Willie, Beverly Sparks, John Macdonald, who helped in researching the history of telegraphy in BC, Tony Rathbone, and Dr. Ken Miki. A very special thanks to Mrs. Patricia Wilson for sharing her knowledge and memories of her grandfather H.O. Bell-Irving and for giving me access to her H.O. Bell-Irving photos and permission to print them.

Bibliography

Callwood, June. *The Naughty Nineties.* Toronto, ON: Natural Science of Canada, 1977.
Campbell, K. Mack. *Cannery Village: Company Town.* Victoria, BC: Trafford, 2004.
Dawson, William Leon. *The Birds of California.* San Diego, CA: South Moulton Co., 1923.
Edwards, Isabel. *Ruffles on my Longjohns.* Surrey, BC: Hancock House, 1980.
Fukawa, Masako. *Nikkei Fishermen on the BC Coast.* Madeira Park, BC: Harbour Publishing, 2007.
George, Chief Earl Maquinna. *Living On The Edge: Nuu-Chah-Nulth History from an Ahousaht Chief's Perspective.* Winlaw, BC: Sononis Press, 2003.
Harris, Douglas C. *Landing Native Fisheries.* Vancouver, BC: UBC Press, 2008.
Hilton, Susanne, and John C. Rath. Eds. *Oowekeeno Oral Traditions as told by the late Chief Simon Walkus Sr.* Ottawa, ON: National Museums of Canada, 1982.
Karliner, Robert, and S.C. Heal. *Stand by—let'er go! The memoirs of a commercial fisherman.* Vancouver, BC: Cordillera Books, 2004.
Knight, Rolf. *Indians at Work.* Vancouver, BC: New Star Books, 1996.
Large, Dr. R.G. *Drums and Scalpel.* Vancouver, BC: Mitchell Press, 1968.
Lyons, Cicely. *Salmon Our Heritage.* Vancouver, BC: Mitchell Press, 1969.
Mack, Clayton. *Grizzlies & White Guys.* Madeira Park, BC: Harbour Publishing, 1993.
McKervill, Hugh M. *The Salmon People.* Sidney, BC: Gray's Publishing, Ltd., 1967.
McKervill, Hugh M. *Darby of Bella Bella.* Toronto, ON: The Ryerson Press, 1964.
Meggs, Geoff. *Salmon: The Decline of the British Columbia Fishery.* Vancouver, BC: Douglas & McIntyre, 1991.

Milner, Marc. *Canada's Navy: The First Century.* Toronto, ON: University of Toronto Press, 1999.

Newell, Dianne. Ed. *The Development of the Pacific Salmon-Canning Industry.* Montreal, QC: McGill–Queen's University Press, 1989.

Perkins, J. David. *The Canadian Submarine Service.* St. Catharines, ON: Vanwell Publishing, 2000.

Petersen, Lester R. "Fishing Rivers Inlet By Sail And Oar," in *Raincoast Chronicles First Five.* Madeira Park, BC: Harbour Publishing, 1976.

Sinclair, Bertrand W. *Poor Man's Rock.* New York: Little, Brown, & Co., 1920.

Stacey, Duncan, and Susan Stacey. *Salmonopolis: The Steveston Story.* Madeira Park, BC: Harbour Publishing, 1994.

Stevenson, David. "The Oowekeeno people: A cultural history." Hull, QC: National Museum of Man History, 1980 (unpublished).

Turner–Turner, J. *Three Years' Hunting and Trapping in America and the Great North-West.* London: Maclure & Co., 1888.

Sources

Adriann Jacob Sem Van Alphen, personal letters
Anglo British Columbia Packing Co. records, UBC Archives
Bell-Irving, H.O., journals, Vancouver Archives
Bell-Irving, Ian, interviewed on tape by Susan Richards de Wit, c. 2001
The British Colonist
The Fisherman
Good Hope Cannery business office records, April 1943–August 1945
H.O. Bell-Irving & Co. records, Vancouver Archives
Hustwick, Alfred. "Wireless Telegraphy in BC," *British Columbia Magazine*, March, 1911
Jones, Robert F. "Last Battle in a Most Foreboding Land," *Sports Illustrated*, July 2, 1973
The New York Times
Leona Taylor and Dorothy Mindenhall, "Index of Historical Victoria Newspapers," *Victoria's Victoria*, www.victoriasvictoria.ca

Personal Interviews

2008–2010

Tony Allard
Andy Anderson
John Anderson
Michael Broughton
David Canning
Mildred (née Lauritsen) Dalton
Harold Dawes
Glen Evans
Inga (née Thompson) Fenwick
George Fisher
Marty Grohn
Bob Hassell
Bill Hobbs
Rolf Hundvik
Darion Jones
Marea (née Abelson) Larsen
Patricia Wilson

Alfred Lauritzen
Astrid (née Gjertson) Mahatoo
Sam Maki
Suzie McArthur
Gerry and Rosemary Miller
Rodney Phillipson
Barb Quinn
Ted Richards
Susan Richards de Wit
Ken Robertson
Richard Straw
Carl Sutter
Spring Sutter